SYLVIA NASAR is the author of the bestselling *A Beautiful Mind*, which won the National Book Critics Circle Award for biography and was turned into a film starring Russell Crowe. She is the John S. and James L. Knight Professor at the Columbia Graduate School of Journalism.

From the reviews of *Grand Pursuit*:

'In *Grand Pursuit*, Sylvia Nasar broke the mould not only in mining the dismal science for enthralling human stories, but braiding them into a sequence of stylish bio-essays. From Marx to Marshall, Keynes to Sen, she demonstrated how revolutions in economic thought shaped the history of their age'   *Independent*, Books of the Year

'[Nasar] tells some wonderful stories about their extraordinary lives'
                        ANDREW HILL, *Financial Times*, Books of the Year

'A wonderful book. *Grand Pursuit* is a history of economics which is full of flesh, bloom, and warmth. The author demonstrates that there is far more to economics than Thomas Carlyle's "dismal science". And she does so with all the style and panache that you would expect from the author of the 1998 bestseller, *A Beautiful Mind*. *Grand Pursuit* deserves a place not only in every economist's study but also on every serious reader's bedside table'   *The Economist*

'I felt deeply stimulated and enriched by Sylvia Nasar's superlative survey of economic history, which tells you why economics arose as a discipline and why it still can't solve most of the conundrums to which we seek answers'   ALAIN DE BOTTON, *New Statesman*, Books of the Year

'Nasar has a story to tell, a story of tragedy, triumph, and economic genius ... [The] book .... wonderful parts ... A rich, in pl........ .... *Book Review*

'This is an utterly fascinating book on many levels ... No lesser mind could have written a book so rich, so compelling, so important, and so much fun'   *Boston Globe*

'Nasar's is a bracingly optimistic, can-do perspective ... she has mastered a formidable body of economic literature, summarized fairly and lucidly ... For anyone wanting a humane, spirited and well-informed guide to how economists contributed to a large stretch of the West's halcyon years, *Grand Pursuit* enjoyably does the job'   DAVID KYNASTON, *TLS*

GRAND
PURSUIT

THE STORY OF THE PEOPLE
WHO MADE MODERN ECONOMICS

SYLVIA NASAR

FOURTH ESTATE • *London*

Fourth Estate
An imprint of HarperCollins*Publishers*
77–85 Fulham Palace Road
Hammersmith, London W6 8JB

This Fourth Estate paperback edition published 2012
1

First published in the United States by Simon & Schuster
First published in Great Britain by Fourth Estate in 2011

A catalogue record for this book is
available from the British Library

ISBN 978-1-84115-456-5

Printed and bound in Great Britain by
Clays Ltd, St Ives plc

MIX
Paper from
responsible sources
FSC
www.fsc.org   FSC™ C007454

FSC™ is a non-profit international organisation established to promote
the responsible management of the world's forests. Products carrying the
FSC label are independently certified to assure customers that they come
from forests that are managed to meet the social, economic and
ecological needs of present and future generations,
and other controlled sources.

Find out more about HarperCollins and the environment at
www.harpercollins.co.uk/green

FOR MY PARENTS

Contents

ACT III: CONFIDENCE

Preface

The Nine Parts of Mankind

The experience of nations with well-being is exceedingly brief.
Nearly all, throughout history, have been very poor.

> *John Kenneth Galbraith*, The Affluent Society, *1958*[1]

In a Misery of this Sort, admitting some few Lenities, and those
too but a few, nine Parts in ten of the whole Race of Mankind
drudge through Life.

> *Edmund Burke*, A Vindication of Natural Society, *1756*[2]

The idea that humanity could turn tables on economic necessity—mastering rather than being enslaved by material circumstances—is so new that Jane Austen never entertained it.

Consider the world of Georgian opulence that the author of *Pride and Prejudice* inhabited. A citizen of a country whose wealth "excited the wonder, the astonishment, and perhaps the envy of the world" her life coincided with the triumphs over superstition, ignorance, and tyranny we call the European Enlightenment.[3] She was born into the "middle ranks" of English society when "middle" meant the opposite of average or typical. Compared to Mr. Bennett in *Pride and Prejudice* or even the unfortunate Ms. Dashwoods in *Sense and Sensibility*,[4] the Austens were quite impecunious. Nonetheless, their income of £210 a year exceeded that of 95 percent of English families at the time.[5] Despite the "vulgar economy" that Austen was required to practice to prevent "discomfort, wretchedness and ruin,"[6] her family owned property, had some leisure, chose their professions, went to school, had books, writing paper, and newspapers at their disposal. Neither Jane nor her sister Cassandra were forced to hire themselves out as governesses—the dreaded fate that awaits Emma's rival Jane—or marry men they did not love.

The gulf between the Austens and the so-called lower orders was, in the words of a biographer, "absolute and unquestioned."[7] Edmund Burke, the philosopher, railed at the plight of miners who "scarce ever see the Light of the Sun; they are buried in the Bowels of the Earth; there they work at a severe and dismal Task, without the least Prospect of being delivered from it; they subsist upon the coarsest and worst sort of Fare; they have their Health miserably impaired, and their Lives cut short."[8] Yet in terms of their standard of living, even these "unhappy wretches" were among the relatively fortunate.

The *typical* Englishman was a farm laborer.[9] According to economic historian Gregory Clark, his material standard of living was not much better than that of an average Roman slave. His cottage consisted of a single dark room shared night and day with wife, children, and livestock. His only source of heat was a smoky wood cooking fire. He owned a single set of clothing. He traveled no farther than his feet could carry him. His only recreations were sex and poaching. He received no medical attention. He was very likely illiterate. His children were put to work watching the cows or scaring the crows until they were old enough to be sent into "service."

In good times, he ate only the coarsest food—wheat and barley in the form of bread or mush. Even potatoes were a luxury beyond his reach. ("They are very well for you gentry but they must be terribly costly to rear," a villager told Austen's mother).[10] Clark estimates that the British farm laborer consumed an average of only 1500 calories a day, one third fewer than a member of a modern hunter-gatherer tribe in New Guinea or the Amazon.[11] In addition to suffering chronic hunger, extreme fluctuations in bread prices put him at risk of outright starvation. Eighteenth-century death rates were extraordinarily sensitive to bad harvests and wartime inflations.[12] Yet the typical Englishman was better off than his French or

German counterpart, and Burke could assure his English readers that this "slavery with all its baseness and horrors that we have at home is nothing compared to what the rest of the world affords of the same Nature."[13]

Resignation ruled. Trade and the Industrial Revolution had swelled Britain's wealth, as the Scottish philosopher Adam Smith predicted in *The Wealth of Nations* in 1776. Still, even the most enlightened observers accepted that these could not trump God's condemnation of the mass of humanity to poverty and "painful toil . . . all the days of your life." Stations in life were ordained by the Deity or nature. When a loyal retainer died, he or she might be praised for "having performed the duties of the Station of life in which he had been pleased to place her in this world."[14] The Georgian reformer Patrick Colquhoun had to preface his radical proposal that the state educate the children of the poor with assurances that he did not mean that they "should be educated in a manner to elevate their minds above the rank they are destined to fill in society" lest "those destined for laborious occupations and an inferior situation in life" become discontented.[15]

In Jane Austen's world everybody knew his or her place, and no one questioned it.

A mere fifty years after her death, that world was altered beyond recognition. It was not only the "extraordinary advance in wealth, luxury and refinement of taste"[16] Or the unprecedented improvement in the circumstances of those whose condition was assumed to be irremediable. The late Victorian statistician Robert Giffen found it necessary to remind his audience that in Austen's day wages had been only half as high and "periodic starvation was, in fact, the condition of the masses of working men throughout the kingdom fifty years ago . . ."[17] It was the sense that what had been fixed and frozen through the ages was becoming fluid. The question was no longer if conditions could change but how much, how fast, and at what cost. It was the sense that the changes were not accidental or a matter of luck, but the result of human intention, will, and knowledge.

The notion that man was a creature of his circumstance, and that those circumstances were not predetermined, immutable, or utterly impervious to human intervention is one of the most radical discoveries of all time. It called into question the existential truth that humanity was subject to the dictates of God and nature. It implied that, given new tools, humanity was ready to take charge of its own destiny. It called for cheer and activity rather than pessimism and resignation. Before 1870 economics was mostly about what you couldn't do. After 1870, it was mostly about what you could do.

"The desire to put mankind into the saddle is the mainspring of most economic study," wrote Alfred Marshall, the father of modern economics. Economic possibilities—as opposed to spiritual, political, or military ones—captured the popular imagination. Victorian intellectuals were obsessed with economics and an extraordinary number aspired to produce a great work in that field. Inspired by advances in natural sciences, they began to fashion a tool for investigating the "very ingenious and very powerful social mechanism" that is creating not just unparalleled material wealth, but a wealth of new opportunities. Ultimately, the new economics transformed the lives of everyone on the planet.

Rather than a history of economic thought, the book in your hands is the story of an idea that was born in the golden age before World War I, challenged in the catastrophic interwar years by two world wars, the rise of totalitarian governments, and a great depression, and was revived in a second golden age in the aftermath of World War II.

Alfred Marshall called modern economics an "Organan," ancient Greek for tool, not a body of truths but an "engine of analysis" useful for discovering truths and, as the term implied, an implement that would never be perfected or completed but would always require improvement, adaptation, innovation. His student John Maynard Keynes called economics an "apparatus of the mind" that, like any other science, was essential for analyzing the modern world and making the most of its possibilities.

I chose protagonists who were instrumental in turning economics into an instrument of mastery. I chose men and women with "cool heads but warm hearts"[18] who helped build Marshall's "engine" and innovated Keynes's "apparatus." I chose figures whose temperaments, experiences and genius led them, in response to their own times and places, to ask new questions and propose new answers. I chose figures that took the story from London in the 1840s around the world, ending in Calcutta at the turn of the twenty-first century. I tried to picture what each of them saw when they looked at their world, and to understand what moved, intrigued, inspired them. All of these thinkers were searching for intellectual tools that could help solve what Keynes called "the political problem of mankind: how to combine three things: economic efficiency, social justice and individual liberty." [19]

As Keynes's first biographer Roy Harrod explained, that protean figure considered the artists, writers, choreographers, and composers he loved and admired to be "the trustees of civilization." He aspired to a humbler but no less necessary role for economic thinkers like himself: to be "the trustees, not of civilization, but of the possibility of civilization." [20]

Thanks in no small part to such trustees, the notion that the nine parts of mankind could free itself from its age-old fate took hold during the Victorian era in London. From there it spread outward like ripples in a pond until it had transformed societies around the globe.

It is still spreading.

Act I

HOPE

Prologue

Mr. Sentiment Versus Scrooge

It was the worst of times.

When Charles Dickens returned from his triumphant American reading tour in June 1842, the specter of hunger was stalking England.[1] The price of bread had doubled after a string of bad harvests. The cities were mobbed by impoverished rural migrants looking for work or, failing that, charity. The cotton industry was in the fourth year of a deep slump, and unemployed factory hands were forced to rely on public relief or private soup kitchens. Thomas Carlyle, the conservative social critic, warned grimly, "With the millions no longer able to live . . . it is too clear that the Nation itself is on the way to suicidal death."[2]

A firm believer in education, civil and religious liberty, and voting rights, Dickens was appalled by the upsurge in class hatred.[3] In August a walkout at a cotton mill turned violent. Within days the dispute had escalated into a nationwide general strike for universal male suffrage, called by leaders of a mass movement for a "People's Charter."[4] The Chartists had taken up the principal cause of middle-class Radicals in Parliament—one man, one vote—into the streets. The Tory government of Prime Minister Robert Peel promptly dispatched red-coated marines to round up the agitators. Rank-and-file strikers began drifting back to their factories, but Carlyle, whose history of the French revolution Dickens read and reread, warned darkly that "revolt, sullen, revengeful humor of revolt against the upper classes . . . is more and more the universal spirit of the lower classes."[5]

In the glittering London drawing rooms where lords and ladies lionized him, Dickens's republican sympathies were as hard to overlook as his garish ties. After running into the thirty-year-old literary sensation for the first time, Carlyle described him patronizingly as "a small compact figure, *very* small," adding cattily that he was "dressed a la D'Orsay rather than well"—which is to say as flash as the notorious *French* count.[6] Carlyle's best friend, the Radical philosopher John Stuart Mill, was reminded of Carlyle's description of a Jacobin revolutionary with "a face of dingy blackguardism radiated by genius."[7] At fashionable midnight suppers the Chartist "uprising" provoked bitter arguments. Carlyle backed the Prime Minister who insisted that harsh measures were necessary to keep radicals from exploiting the situation and that the truly needy were already getting help. Dickens, who swore that he "would go farther at all times to see Carlyle than any man alive,"[8] nonetheless maintained that prudence and justice both demanded that the government grant relief to the able-bodied unemployed and their families.

The Hungry Forties revived a debate that had raged during the famine years, 1799 to 1815, of the Napoleonic Wars. At issue was the controversial law of population propounded by the Reverend Thomas Robert Malthus. A contemporary of Jane Austen and England's first professor of political economy, Malthus was a shy, softhearted Church of England clergyman with a harelip and a hard-edged mathematical mind. While still a curate, he had been tormented by the hunger in his rural parish. The Bible blamed the innate sinfulness of the poor. Fashionable French philosophers like his father's friend the Marquis de Condorcet blamed the selfishness of the rich. Malthus found neither explanation compelling and felt bound to search for a better one. *An Essay on the Principle of Population,* published first in 1798 and five more times before his death in 1834, inspired Charles Darwin and the other founders of evolutionary theory and prompted Carlyle to dismiss economics as the "dismal science."[9]

The fact that Malthus sought to explain was that, in all societies and all epochs including his own, "nine parts in ten of the whole race of man-kind" were condemned to lives of abject poverty and grinding toil.[10] When not actually starving, the typical inhabitant of the planet lived in chronic fear of death by hunger. There were prosperous years and lean ones, richer and poorer regions, yet the standard of life never departed for long from subsistence.

In attempting to answer the age-old question "Why?" the mild-mannered minister anticipated not only Darwin but Freud. Sex, he argued, was to blame. Whether from observing the wretched lives of his parishion-ers, the influence of natural scientists who were beginning to regard man as an animal, or the arrival of his seventh child, Malthus had concluded that the drive to reproduce trumped all other human instincts and abili-ties, including rationality, ingenuity, creativity, even religious belief.

From this single provocative premise, Malthus deduced the principle that human populations tended always and everywhere to grow faster than the food supply. His reasoning was deceptively simple: Picture a situation in which the supply of food is adequate to sustain a given population. That happy balance can't last any more than could Adam and Eve's tenure in paradise. Animal passion drives men and women to marry sooner and have bigger families. The food supply, meanwhile, is more or less fixed in all but the very long run. Result: the amount of grain and other staples that had just sufficed to keep everyone alive would no longer be enough. Inevitably, Malthus concluded, "the poor consequently must live much worse."[11]

In any economy where businesses compete for customers and workers for jobs, an expanding population meant more households contending for the food supply, and more workers competing for jobs. Competition would drive down wages while simultaneously pushing food prices higher. The average standard of living—the amount of food and other necessities available for each person—would fall.

At some point, grain would become so expensive and labor so cheap that the dynamic would reverse itself. As living standards declined, men and women would once again be forced to postpone marriage and have fewer children. A shrinking population would mean falling food prices as fewer households competed for the available food. Wages would rise as fewer workers competed for jobs. Eventually, as the food supply and population moved back into balance, living standards would creep back to their old level. That is, unless Nature's "great army of destruction"[12]—war, disease, and famine—intervened to hurry the process, as happened, for example, in the fourteenth century, when the Black Plague wiped out millions, leaving behind a smaller population relative to the output of food.

Tragically, the new balance would prove no more durable than the original one. "No sooner is the laboring class comfortable again," Malthus wrote sadly, "than the same retrograde and progressive movements with respect to happiness are repeated."[13] Trying to raise the average standard of living is like Sisyphus trying to roll his rock to the top of the hill. The faster Sisyphus gets almost there, the sooner he triggers the reaction that sends the boulder tumbling down the slope again.

Attempts to flout the law of population were doomed. Workers who held out for above-market wages wouldn't find jobs. Employers who paid their workers more than their competitors did would lose their customers as higher labor costs forced them to raise prices.

For Victorians, the most objectionable implication of Malthus's law was that charity might actually increase the suffering it was intended to ease—a direct challenge to Christ's injunction to "love thy neighbor as thyself."[14] In fact, Malthus was extremely critical of the traditional English welfare system, which provided relief with few strings attached, for rewarding the idle at the expense of the industrious. Relief was proportional to family size, in effect encouraging early marriage and large families. Conservative and liberal taxpayers alike found Malthus's argument so persuasive that Parliament passed, virtually without opposition, a new Poor Law in 1834 that effectively restricted public relief to those who agreed to become inmates of parish workhouses.

"Please, sir, I want some more." As Oliver Twist discovers after making his famous plea, workhouses were essentially prisons where men and women were segregated, put to work at unpleasant tasks, and subjected to harsh discipline—all in return for a place to sleep and "three meals of thin gruel a day, with an onion twice a week, and half a roll on Sundays."[15] The fare in most workhouses probably wasn't as meager as the starvation diet Dickens described in his novel, but there is no doubt that these institutions topped the list of working-class grievances.[16] Like most reform-minded middle-class liberals, Dickens considered the new Poor Law morally repulsive and politically suicidal and the theory on which it was based a relic of a barbaric past. He had recently returned from America with its "thousands of millions of acres of land yet unsettled and uncleared" and where the inhabitants were in "the custom of hastily swallowing large quantities of animal food, three times a-day,"[17] and found the notion that abolishing the workhouse would cause the world to run out of food absurd.

Bent on striking a blow for the poor, Dickens began early in 1843 to write a tale about a rich miser's change of heart, a tale that he liked to think of as a sledgehammer capable of "twenty times the force—twenty thousand times the force" of a political pamphlet.[18]

A Christmas Carol, argues the economic historian James Henderson, is an attack on Malthus.[19] The novel is bursting with delicious smells and tastes. Instead of a rocky, barren, overpopulated island where food is scarce, the England of Dickens's story is a vast Fortnum & Mason where the shelves are overflowing, the bins are bottomless, and the barrels never run dry. The Ghost of Christmas Past appears to Scrooge perched on a "kind of throne," with heaps of "turkeys, geese, game, poultry, brawn, great joints of meat, sucking-pigs, long wreaths of sausages, mince-pies, plum-puddings, barrels of oysters, red-hot chestnuts, cherry-cheeked apples, juicy oranges, luscious pears, immense twelfth-cakes, and seething bowls of punch, that made the chamber dim with their delicious steam." "Radiant" grocers, poulterers, and fruit and vegetable dealers invite Londoners into their shops to inspect luscious "pageants" of food and drink.[20]

In an England characterized by New World abundance rather than Old World scarcity, the bony, barren, anorexic Ebenezer Scrooge is an anachronism. As Henderson observes, the businessman is "as oblivious to the new spirit of human sympathy as he is to the bounty with which he is surrounded."[21] He is a diehard supporter of the treadmill and workhouse literally and figuratively. "They cost enough," he insists, "and those who are badly off must go there." When the Ghost of Christmas Past objects that "many can't go there; and many would rather die," Scrooge says coldly, "If they would rather die, they had better do it, and decrease the surplus population."

Happily, Scrooge's flinty nature turns out to be no more set in stone than the world's food supply is fixed. When Scrooge learns that Tiny Tim is one of the "surplus" population, he recoils in horror at the implications of his old-fashioned Malthusian religion. "No, no," he cries, begging the Spirit to spare the little boy. "What then?" the Spirit replies mockingly. "If he be like to die, he had better do it, and decrease the surplus population." [22] Scrooge repents, resolves to give his long-suffering clerk, Bob Cratchit, a raise, and sends him a prize turkey for Christmas. By accepting the more hopeful, less fatalistic view of Dickens's generation in time to alter the course of future events, Scrooge refutes the grim Malthusian premise that "the blind and brutal past" is destined to keep repeating itself.

The Cratchits' joyous Christmas dinner is Dickens's direct riposte to Malthus, who uses a parable about "Nature's mighty feast" to warn of the unintended consequences of well-meaning charity. A man with no means of support asks the guests to make room for him at the table. In the past, the diners would have turned him away. Beguiled by utopian French theories, they decide to ignore the fact that there is only enough food for the invited guests. They fail to foresee when they let the newcomer join them that more gatecrashers will arrive, the food will run out before everyone has been served, and the invited guests' enjoyment of the meal will be "destroyed by the spectacle of misery and dependence." [23]

The Cratchits' groaning board, wreathed with the family's beaming faces, is the antithesis of Malthus's tense, tightly rationed meal. In contrast to Nature's grudging portions, there is Mrs. Cratchit's pudding—"like a speckled cannon-ball, so hard and firm, blazing in half of half-a-quartern of ignited brandy, and bedecked with Christmas holly stuck into the top"—not large enough for seconds perhaps, but ample for her family. "Mrs. Cratchit said that now the weight was off her mind, she would confess she had had her doubts about the quantity of flour. Everybody had something to say about it, but nobody said or thought it was at all a small pudding for a large family. It would have been flat heresy to do so. Any Cratchit would have blushed to hint at such a thing."[24]

The Christmas spirit was catching. By the story's end, Scrooge had even stopped starving himself. Instead of slurping his customary bowl of gruel in solitude, the new Scrooge surprises his nephew by showing up unannounced for Christmas dinner. Needless to say, his heir hastens to set a place for him at the table.

Dickens's hope that *A Christmas Carol* would strike the public like a sledgehammer was fulfilled. Six thousand copies of the novel were sold between the publication date of December 19 and Christmas Eve, and the tale would stay in print for the rest of Dickens's life—and ever since.[25] Dickens's depiction of the poor earned him satirical labels such as "Mr. Sentiment,"[26] but the novelist never wavered in his conviction that there was a way to improve the lot of the poor without overturning existing society.

Dickens was too much a man of business to imagine that schemes for bettering social conditions could succeed unless they could be paid for. He was a "pure modernist" and "believer in Progress" rather than an opponent of the Industrial Revolution. Wildly successful while still in his twenties, he had gone too far on his own talent to doubt that human ingenuity was climbing into the driver's seat. Having escaped poverty by making his way in the new mass-media industry, Dickens was impatient with conservatives such as Carlyle and socialists such as Mill who refused to admit that, as a society, "we have risen slowly, painfully, and with many a hard struggle out of all this social degradation and ignorance" and who "look back to all this blind and brutal past with an admiration they will not grant to the present."[27]

Dickens's sense that English society was waking up, as if from a long nightmare, proved prescient. Within a year of the Chartist "uprising," a new mood of tolerance and optimism was palpable. The Tory prime minister admitted privately that many of the Chartists' grievances were justified.[28] Labor leaders rejected calls for class warfare and backed employers' campaign to repeal import duties on grain and other foodstuffs. Liberal politicians responded to parliamentary commissions on child labor, industrial accidents, and other evils by introducing the Factory Acts of 1844, legislation regulating the hours of women and children.

Dickens never imagined that the world could get along without the calculating science of economics. Instead, he hoped to convert political economists as the Ghost of Christmas Future had converted Scrooge. He wanted them to stop treating poverty as a natural phenomenon, assuming that ideas and intentions were of no importance, or taking for granted that the interests of different classes were diametrically opposed. Dickens was especially eager for political economists to practice "mutual explanation, forbearance and consideration; something . . . not exactly stateable in figures."[29] When he launched his popular weekly, *Household Words,* he did so with a plea to economists to humanize their discipline. As he wrote in his inaugural essay, "Political economy is a mere skeleton unless it has a little human covering, and filling out, a little human bloom upon it, and a little human warmth in it."[30]

Dickens was not alone. There were—and would be—men and women in London and all over the world who reached the same conclusion. Having overcome formidable obstacles, they too saw man as a creature of circumstance. They too realized that the material conditions of life for the "nine parts in ten of the whole race of mankind" were no longer immutable, predetermined by the "blind and brutal past," and utterly beyond human control or influence. Convinced that economic circumstances were open to human intervention yet skeptical of utopian schemes and "artificial societies" imposed by radical elites, they devoted themselves to fashioning an "engine of analysis"[31] (or, as a later economist put it, an "apparatus of the mind")[32] that they could use to understand how the modern world worked and how humanity's material condition—on which its moral, emotional, intellectual, and creative condition depended—could be improved.

Chapter I

Perfectly New: Engels and Marx in the Age of Miracles

The exact point is that it has not gone on a long time. [It is] perfectly new. . . .

Our system though curious and peculiar may be worked safely . . . if we wish to work it, we must study it.

—*Walter Bagehot,* Lombard Street [1]

"See to it that the material you've collected is soon launched into the world," the twenty-three-year-old Friedrich Engels wrote to his corevolutionist, Karl Marx. "It's high time. Down to work, then, and quickly into print!" [2]

In October 1844, continental Europe was a smoldering volcano threatening to erupt. Marx, the son-in-law of a Prussian nobleman and editor of a radical philosophy journal, was in Paris, where he was supposed to be writing an economic treatise to prove with mathematical certainty that revolution must come. Engels, the scion of prosperous Rhenish textile merchants, was at his family's estate, up to his eyebrows in English newspapers and books. He was drafting a "fine bill of indictment" against the class to which he and Marx belonged.[3] His only anxiety was that the revolution would arrive before the galleys.

A romantic rebel with literary aspirations, Engels was already an "embryonic revolutionary" and "enthusiastic communist" when he met Marx for the first time two years earlier. Having spent his adolescence freeing himself from his family's strict Calvinism, the slender, fair, severely near-sighted Royal Prussian artillerist had trained his sights on the twin tyrannies of God and Mammon. Convinced that private property was the root of all evil and that a social revolution was the only way to establish a just society, Engels had yearned to live the "true" life of a philosopher. To his infinite regret, he was predestined for the family trade. "I am not a Doctor," he had corrected the wealthy publisher of a radical newspaper who mistook him for a scholar, adding that he could "never become one. I am only a businessman."[4]

Engels Senior, a fervid Evangelical who clashed frequently with his freethinking son, wouldn't have it any other way. As a proprietor, he was quite progressive. He supported free trade, adopted the latest British spinning equipment in his factory in the Wuppertal, and had recently opened a second plant in Manchester, the Silicon Valley of the industrial revolution. But as a father he could not stomach the notion of his eldest son and heir as a professional agitator and freelance journalist. When the global cotton trade collapsed in the spring of 1842, followed by the Chartist strikes, he insisted that the young Engels report to work at Ermen & Engels in Manchester as soon his compulsory military service was over.

Bowing to filial duty hardly meant the death of Engels's dream of becoming the scourge of authority in all forms. Manchester was notorious for the militancy of its factory hands. Convinced that the industrial strife was a prelude to wider insurrection, Engels had been only too delighted to go where the action was and to use the opportunity to advance his writing career.

En route to England in November, he had stopped in Cologne, where he visited the grubby offices of the prodemocracy newspaper *Rheinische Zeitung,* to which he had been contributing occasional articles under the byline *X.* The new editor was a brusque, cigar-smoking, exceedingly myopic philosopher from Trier who treated him rudely. Engels had taken no offense and had been rewarded with an assignment to report on the prospects for revolution in England.

· · ·

When Engels arrived in Manchester, the general strike had petered out and the troops had returned to their London barracks, but there were unemployed men hanging around street corners, and many of the mills were still idle. Despite his conviction that the factory owners would rather see their employees starve than pay a living wage, Engels could not help noticing that the English factory worker ate a great deal better than his counterpart in Germany. While a worker at his family's textile mill in Barmen dined almost exclusively on bread and potatoes, "Here he eats beef every day and gets a more nourishing joint for his money than the richest man in Germany. He drinks tea twice a day and still has enough money left over to be able to drink a glass of porter at midday and brandy and water in the evening." [5]

To be sure, unemployed cotton workers had had to turn to the Poor Law and private soup kitchens to avoid "absolute starvation," and Edwin Chadwick's just-published *Report on the Sanitary Condition of the Labouring Population of Great Britain* revealed that the average male life span in Manchester was seventeen years, half that of nearby rural villages, and that just one in two babies survived past age five. Chadwick's graphic descriptions of streets that served as sewers, cottages damp with mold, rotting food, and rampant drunkenness demonstrated that British workers had ample grounds for resentment. [6] But while Carlyle, the only Englishman Engels admired, warned of working-class revolt, Engels found that most middle-class Englishmen considered the possibility remote and looked to the future with "remarkable calm and confidence." [7]

Once settled in his new home, Engels resolved the conflict between his family's demands and his revolutionary ambitions in a characteristically Victorian fashion. He lived a double life. At the office and among his fellow capitalists, he resembled the "sprightly, good humored, pleasant" Frank Cheeryble, the "nephew of the firm" in Dickens's *Nicholas Nickleby* who "was coming to take a share in the business here" after "superintending it in Germany for four years." [8] Like the novel's attractive young businessman, Engels dressed impeccably, joined several clubs, gave good dinners, and kept his own horse so that he could go fox hunting at friends' estates. In his other, "true" life, he "forsook the company and the dinner-parties, the port-wine and champagne" to moonlight as a Chartist organizer and investigative journalist.[9] Inspired by the exposés of English reformers and often accompanied by an illiterate Irish factory girl with whom he was having an affair, Engels spent his free time getting to know Manchester "as intimately as my native town," gathering materials for the dramatic columns and essays he filed to various radical newspapers.

Engel's twenty-one months as a management trainee in England led him to discover economics. While German intellectuals were obsessed by religion, the English seemed to turn every political or cultural issue into an economic question. It was especially true in Manchester, a stronghold of English political economy, the Liberal Party, and the Anti–Corn Law League. To Engels, the city represented the interconnections between the industrial revolution, working class militancy, and the doctrine of laissez-faire. Here "it was forcibly brought to my notice that economic factors, hitherto ignored or at least underestimated by historians, play a decisive role in the development of the modern world," he later re-called.[10]

Frustrated as he was by his lack of a university education, particularly his ignorance of the works of Adam Smith, Thomas Malthus, David Ricardo, and other British political economists, Engels was nonetheless perfectly confident that British economics was deeply flawed. In one of the last essays he wrote before leaving England, he hastily roughed out the essential elements of a rival doctrine. Modestly, he called this fledgling effort "Outlines of a Critique of Political Economy." [11]

Across the English Channel in St. Germain-en-Laye, the wealthiest suburb of Paris, Karl Marx had buried himself in histories of the French Revolution. When Engels's final piece arrived in the post, he was jolted back to the present, electrified by his correspondent's "brilliant sketch on the critique of economic categories." [12]

Marx too was the prodigal (and profligate) son of a bourgeois father. He too was an intellectual who felt trapped in a philistine age. He shared Engels's sense of German intellectual and cultural superiority, admired all things French, and bitterly resented British wealth and power. Yet he was in many ways Engels's opposite. Domineering, impetuous, earnest, and learned, Marx had none of the other man's glibness, adaptability, or cheerful bonhomie. Only two and a half years older, Marx was not only married and the father of a baby girl but also a doctor of philosophy who insisted on being addressed as such. A short, powerfully built, almost Napoleonic figure, he had thick jet-black hair that sprouted from cheeks, arms, nose, and ears. His "eyes glowed with an intelligent and malicious fire," and, as his assistant at the *Rheinische Zeitung* recalled, his favorite conversation starter was "I am going to annihilate you." [13] One of his biographers, Isaiah Berlin, identified Marx's "belief in himself and his own powers" as his "single most outstanding characteristic." [14]

While Engels was practical and efficient, Marx was, as George Bernard Shaw pointed out, "without administrative experience" or any "business contact with a living human being." [15] He was undeniably brilliant and erudite, but he had never acquired Engels's work ethic. Whereas Engels was ready at any hour to roll up his sleeves and start writing, Marx was more likely to be found in a café, drinking wine and arguing with Russian aristocrats, German poets, and French socialists. As one of his backers once reported, "He reads a lot. He works with extraordinary intensity . . . He never finishes anything. He interrupts every bit of research to plunge into a fresh ocean of books . . . He is more excitable and violent than ever, especially when his work has made him ill and he has not been to bed for three or four nights on end." [16]

Marx had been forced to turn to journalism when he failed to obtain an academic post at a German university and his long-suffering family finally cut him off financially. [17] After just six months at his newspaper job in Cologne—"the very air here turns one into a serf"—he picked a fight with the Prussian censor and quit. Luckily, Marx was able to convince a wealthy Socialist to finance a new philosophical journal, the *Franco-German Annals,* and appoint him to run it in his favorite city, Paris.

Engels's reports from Manchester emphasizing the link between economic causes and political effects made a powerful impression on Marx. Economics was new to him. The terms *proletariat, working class, material conditions,* and *political economy* had yet to crop up in his correspondence.

As his letter to his patron shows, he had envisioned an alliance of "the ene-
mies of philistinism, i.e. all thinking and suffering people," but his goal was
reforming consciousness, not abolishing private property. His contribution
to the first and only issue of the *Franco-German Annals* makes clear that
Marx meant to hurl criticisms, not paving stones, at the powers that be:
"Every individual must admit to himself that he has no precise idea about
what ought to happen. However, this very defect turns to the advantage of
the new movement, for it means that we do not anticipate the world with
our dogmas but instead attempt to discover the new world through the
critique of the old."

He went on, "We shall simply show the world why it is struggling . . .
Our program must be: the reform of consciousness . . . the self-
clarification . . . of the struggles and wishes of the age." The philosopher's
role was like that of a priest: "What is needed above all is a *confession,* and
nothing more than that. To obtain forgiveness for its sins, mankind needs
only to declare them for what they are."

Marx and Engels had their first real encounter in August 1844 at the
Café de Regence. Engels stopped in Paris on his way home to Germany expressly to see the man who had earlier rebuffed him. They talked, argued,
and drank for ten straight days, discovering again and again that each had
been thinking the other's thoughts. Marx shared Engels's conviction of
the utter hopelessness of reforming modern society, and the need to free
Germany from God and traditional authority. Engels introduced him to
the idea of the proletariat. Marx felt an immediate sense of identification
with that class. He saw the proletariat not only, as one might expect, as
the "naturally arising poor" but also as the "artificially impoverished . . .
masses resulting from the drastic dissolution of society,"[18] aristocrats who
had lost their lands, bankrupt businessmen, and unemployed academics.

Like Carlyle and Engels, Marx seized on hunger and rebelliousness
as evidence of the bourgeoisie's unfitness for rule: "absolutely imperative
need" will drive the proletariat to overthrow its oppressors, he predicted.[19]
By abolishing private property, the proletariat would free not only itself
but the entire society. As the historian Gertrude Himmelfarb observes,
Engels and Marx were hardly the only Victorians who were convinced that

modern society was suffering from a terminal illness.[20] They differed from Carlyle and other social critics chiefly in their emphasis on the inevitability of the demise of the existing social order. Even as they struggled to free themselves from Protestant dogma, they became convinced that the economic collapse and violent revolution they predicted were fates from which there was no escape—so to speak, predestined. While Carlyle's doomsday message was meant to inspire repentance and reform, theirs was meant to urge their readers to get on the right side of history before it was too late.

In *The Condition of the Working Class in England in 1844*, Engels had made a compelling, if not necessarily accurate, case that England's industrial workforce normally lived in a state of semistarvation and that famine had driven it to violence against factory owners in 1842. What his journalistic account could not establish was that workers' precarious existence was immutable and that no solution existed short of the overthrow of English society and the imposition of a Chartist dictatorship. This is the argument that Engels had kept losing with his English acquaintances and the problem he had urged Marx to take up. He explained to Marx that in England, social and moral problems were being redefined as economic problems, and social critics were being forced to grapple with *economic* realities. Just as the disciples of the German philosopher Georg Hegel had used religion to dethrone religion and expose the hypocrisy of Germany's ruling elite, they would have to use the principles of political economy to eviscerate the hateful English "religion of money."

When the new friends parted, Engels went home to Germany to pour out his charges of "murder, robbery and other crimes on a massive scale" against the British business class (and, by implication, Germany's as well).[21] Working in his family's cotton thread factory had confirmed Engels's feeling that business was "filthy."[22] He had "never seen a class so deeply demoralized, so incurably debased by selfishness, so corroded within, so incapable of progress, as the English bourgeoisie." These "bartering Jews," as he called the businessmen of Manchester, were devotees of "Political Economy, the Science of Wealth," indifferent to the suffering of their workers as long as they made a profit and, indeed, to all human values except money. "The huckstering spirit" of the English upper classes was as repugnant as the "Pharisaic philanthropy" that they dispensed to the poor after "sucking out their very life-blood." With English society increasingly "divided into millionaires and paupers," the imminent "war of the poor against the rich" would be "the bloodiest ever waged."[23] As fast and fluent a writer as he was a talker, Engels finished his manuscript in less than twelve weeks.

All the while, Engels badgered Marx to "Do try and finish your political economy book . . . It must come out soon."[24] His own book was published in Leipzig in July 1845. *The Condition of the Working Class in England* drew favorable reviews and sold well even before the economic and political crises that the author correctly forecast for "1846 or 1847" gave it the added cache of successful prophecy. *Das Kapital,* the grandiose treatise in which Marx promised to reveal the "law of motion of modern society," took twenty years longer.[25]

In 1849, when Henry Mayhew, a *London Morning Chronicle* correspondent, climbed to the Golden Gallery atop St. Paul's Cathedral to get a bird's-eye view of his hometown, he found that "it was impossible to tell where the sky ended and the city began."[26] At nearly 20 percent a decade, the city's growth "seemed to obey no known law."[27] By the middle of the century, the population had swelled to two and one half million. There were more than enough Londoners to populate two Parises, five Viennas, or the eight next-largest English cities combined.[28]

London "epitomized the 19th century economic miracle."[29] The pool of London was the world's biggest and most efficient port. As early as 1833, a partner in the Barings Brothers Bank observed that London had become the "center upon which commerce must turn." London's wet docks covered hundreds of acres and had become a prime tourist attraction—not least because of a twelve-acre underground wine cellar that gave visitors a chance to taste the Bordeaux. The smells—pungent tobacco, overpowering rum, sickening hides and horn, fragrant coffee and spices—evoked a vast global trade, an endless stream of migrants, and a far-flung empire.

"I know nothing more imposing than the view which the Thames offers during the ascent from the sea to London Bridge," Engels had confessed in 1842 after seeing London for the first time. "The masses of buildings, the wharves on both sides, especially from Woolwich upwards, the countless ships along both shores, crowding ever closer and closer together, until, at last, only a narrow passage remains in the middle of the river, a passage through which hundreds of steamers shoot by one another; all this is so vast, so impressive, that a man cannot collect himself." [30]

London's railway stations were "vaster than the walls of Babylon ... vaster than the temple of Ephesus," John Ruskin, the art historian, claimed. "Night and day the conquering engines rumbled," wrote Dickens in *Dombey and Son*. From London, a traveler could go as far north as Scotland, as far east as Moscow, as far south as Baghdad. Meanwhile, the railroad was pushing London's boundaries ever farther into the surrounding countryside. As Dickens related, "The miserable waste ground, where the refuse-matter had been heaped of yore, was swallowed up and gone, and in its frowsy stead were tiers of warehouses, crammed with rich goods and costly merchandise. Bridges that had led to nothing, led to villas, gardens, churches, healthy public walks. The carcasses of houses, and beginnings of new thoroughfares, had started off upon the line at steam's own speed, and shot away into the country in a monster train." [31]

The financial heart of world commerce beat in the "City," London's financial center. The financier Nathan Mayer Rothschild, not given to exaggeration, called London "the bank of the world." [32] Merchants came there to raise short-term loans to finance their global trade, and governments floated bonds to build roads, canals, and railways. Although the London stock exchange was still in its infancy, the City's merchants and bill discounters attracted three times the amount of "borrowable money" as New York and ten times as much as Paris.[33] Bankers', investors', and merchants' hunger for information helped make London into the world's media and communications center. "Anyone can get the news," a Rothschild complained in 1851 when the advent of the telegraph made his carrier pigeon network obsolete.[34]

London, not the new industrial towns in the north, boasted the big-

gest concentration of industry in the world, employing one in six manufacturing workers in England, nearly half a million men and women.[35]
That was roughly ten times the number of cotton workers in Manchester.
The "dark satanic mills" in William Blake's *Jerusalem* probably weren't
in the Coketowns of northern England. Like the monster Albion flour
mill, which employed five hundred workers and was powered by one
of James Watt's gargantuan steam engines, they were more likely on the
Thames in London.[36] A popular 1850s travel guide refers to "water works,
gas works, shipyards, tanning yards, breweries, distilleries, glass works the
extent of which would excite no little surprise in those who for the first
time visited them."[37] True, London had no single dominant industry such
as textiles, and most of its manufacturing firms employed fewer than ten
hands,[38] but entire industries—printing in Fleet Street, paint, precision
instruments in Camden, and furniture making around Tottenham Road—
were concentrated in London. The vast shipyards at Poplar and Millwall
employed fifteen thousand men and boys to build the biggest steamships
and armor-plated warships afloat. But while factory towns like Leeds and
Newcastle supplied the bulk of England's exports, most of London's man-
ufacturers catered to the needs of the city itself. Wandsworth had its flour
mills, Whitechapel its sugar refiners, Cheapside its breweries, Smithfield
its cattle markets, and Bermondsey its tanneries, candle and soap makers.
Mayhew called London the world's "busiest hive."[39]

Above all, London was the world's biggest market. Here one could get
"at a low cost and with the least trouble, conveniences, comforts, and ame-
nities beyond the compass of the richest and most powerful monarchs."[40]
In the prosperous West End of London "everything shines more or less,
from the window panes to the dog collars" and "the air is colored, almost
scented, by the presence of the biggest society on earth."[41] Regent Street
displayed the greatest collection of "watchmakers, haberdashers, and pho-
tographers; fancy stationer, fancy hosiers, and fancy stay makers; music
shops, shawl shops, jewelers, French glove shops, perfumery, and point
lace shops, confectioners and milliners" the world had ever seen.[42]

Mayhew astutely attributed "the immensity of . . . commerce" in the
city to "the unparalleled prevalence of merchant people in London, and

the consequent vastness of wealth."[43] The *Economist* boasted, "The richest persons in the Empire throng to her. Her scale of living is most magnificent; her rents highest; her opportunities of money-making widest."[44] One in six Britons lived in London, but London accounted for an even bigger share of national income. Incomes were, on average, 40 percent higher than in other English cities, not only because London had more wealthy residents but also because London wages were at least one-third higher than elsewhere. Her huge population and vast income made London by far the greatest concentration of consumer demand in the world. The economic historian Harold Perkin argues that "Consumer demand was the ultimate economic key to the Industrial Revolution," providing a more powerful impetus than the invention of the steam engine or the loom.[45] London's needs, passion for novelty, and growing spending power supplied entrepreneurs with compelling incentives to adopt new technologies and create new industries.

If London attracted some of the richest individuals on earth, it was also a magnet for a large number of the poorest. When Mayhew referred to "the unprecedented multitude of individuals attracted by such wealth to the spot," he meant not only the shopkeepers, tradesmen, lawyers, and doctors who catered specifically to the rich, but also the legions of unskilled migrants from the surrounding rural counties who came to work as servants, seamstresses, shoemakers, carpenters, dockhands, casual laborers, and messengers, or, failing that, as petty criminals, scavengers, and prostitutes.[46] The juxtaposition between rich and poor was rendered more striking by the exodus of the middle classes to the suburbs and, more significant in the minds of observers, by the universal assumption that London presaged the future of society. Poverty was not, of course, new. But in the country, hunger, cold, disease, and ignorance appeared to be the work of nature. In the great capital of the world, misery seemed to be man-made, almost gratuitous. Wasn't the means to relieve it at hand, actually visible in the form of elegant mansions, elaborate gowns, handsome carriages, and lavish entertainments? Well, no. It only looked that way to unsophisticated observers who had no idea that letting the poor eat cake for a day or two would hardly solve the problem of producing enough bread, clothing, fuel, housing, education, and medical care to raise most Englishmen out of poverty. Mayhew was not alone in naïvely supposing that the rows of brick warehouses, "vast emporia," contained wealth "enough, one would fancy, to enrich the people of the whole globe."[47]

Journalists, artists, novelists, social reformers, clergymen, and other students of society were drawn to London as "an epitome of the round world" where "there is nothing one cannot study at first hand."[48] They came there to see where society was heading. While eighteenth-century visitors were apt to focus on sin, crime, and filth, those who flocked to Victorian London were more often struck by its extremes of poverty and wealth.

November was the worst month for air quality in the world's biggest and richest metropolis, observed Charles Dickens in *Bleak House*.[49] On the twenty-ninth of that month in 1847, Friedrich Engels and Karl Marx struggled up Great Windmill Street toward Piccadilly, heads bowed and trying their best to avoid slipping in the ankle-deep mud or being trampled by the human throng. Their extreme myopia and the sulfurous yellow London fog obscured everything more than a foot ahead.

Engels, still as erect as a cadet, and Marx, still with a jet-black mane and magnificent whiskers, were in London to attend a congress of the Communist League, one of many tiny groups comprised of Central European utopians, Socialists, and anarchists, as well as the odd Chartist and occasional Cockney clerk in favor of male suffrage, that flourished in the relative safety of English civil liberties and lenient immigration law. When the recent collapse of a railroad boom spread financial panic in London and on the Continent, the league had hastily convened a meeting to hammer out its hitherto somewhat nebulous goals. Engels had already convinced the league to drop its insipid slogan, "All Men Are Brothers," in favor of the more muscular "Proletarians of All Countries Unite!" He had composed two drafts of a manifesto that he and Marx meant for the league to adopt. They had discussed how they could shoulder aside those in the leadership who were convinced that workers' grievances could be addressed without overthrowing the existing order. "This time we shall have our way," Engels had sworn in his most recent letter to Marx.[50]

They finally found their way to Soho and the Red Lion pub. The headquarters of the German Workers' Educational Union, a front for the illegal league, was on the second floor. The room had a few wooden tables and chairs and, in one corner, a grand piano meant to make refugees from Berlin and Vienna stranded in "unmusical" London feel at home.[51] The air smelled of wet woolens, penny tobacco, and warm beer. For ten days, Engels and Marx dominated the proceedings, navigating the atmosphere of conspiracy and suspicion like fishes in water.

At one point, Marx read Engels's draft manifesto out loud. One delegate recalled the philosopher's relentless logic as well as the "sarcastic curl" of his mouth. Another remembered that Marx spoke with a lisp, which caused some listeners to hear "eight-leaved clovers" when he actually said "workers." [52] Some delegates repudiated Engels and Marx as "bourgeois intellectuals." At the end of the ten days, however, "all opposition . . . was overcome."

The congress voted to adopt their manifesto and agreed to declare itself in favor of "the overthrow of the bourgeoisie, the abolition of private property, and the elimination of inheritance rights." Marx, who had already burned through several family bequests but was, as usual, broke again, was commissioned to draft the final version of the league's call to arms.[53]

Engels had wanted the pamphlet to be a "simple historical narrative" and proposed that it be called *The Communist Manifesto*. He thought it important to tell the story of modern society's origins in order to show why it was destined to self-destruct. He envisioned the *Manifesto* as a sort of Genesis and Revelation rolled into one.[54]

Three years after Engels introduced Marx to English political economy, Marx was already calling himself an economist.[55] He had also absorbed the evolutionary theories that were beginning to pervade the sciences. Like other left-wing disciples of Hegel, he viewed society as an evolving organism rather than one that merely reproduced itself from one generation to another.[56]

He wanted to show that the industrial revolution signified more than the adoption of new technologies and the spectacular leap in production. It had created huge cities, factories, and transport networks. It had launched a vast global trade that made universal interdependence, not national self-sufficiency, the rule. It had imposed new patterns of boom and bust on economic activity. It had torn old social groups from their moorings and created entirely new ones, from millionaire industrialists to poverty-stricken urban laborers.

For a dozen centuries, as empires rose and fell and the wealth of nations waxed and waned, the earth's thin and scattered population had grown by tiny increments. What remained essentially unchanged were man's material circumstances, circumstances that guaranteed that life would remain miserable for the vast majority. Within two or three generations, the industrial revolution demonstrated that the wealth of a nation could grow by multiples rather than percentages. It had challenged the most basic premise of human existence: man's subservience to nature and its harsh dictates. Prometheus stole fire from the gods, but the industrial revolution encouraged man to seize the controls.

Engels and Marx perceived more clearly than most of their contemporaries the newness of the society in which they came of age, and tried to work out its implications more obsessively. Modern society was evolving faster than any society in the past, they believed. The consciousness of change and changeability was a breach in the firmament of traditional truths and received wisdom. In Marx's memorable phrase, "All that is solid melts into air."[57] Surely the vividness of their perceptions owes something to the fact that they came to England as foreign correspondents, so to speak, and that they came from a country that had yet to go through its industrial revolution. The trips from Trier and Barmen in Germany to London were journeys forward in time. Hardly anyone, except perhaps Charles Dickens, was as simultaneously thrilled and revolted by what they witnessed. They professed to despise England's "philistine" commercial culture while envying her wealth and power. Their observations convinced them that in the modern world, political power grew not from the barrels of guns but out of a nation's economic superiority and the energy of its business class.

England was the colossus astride the modern world. "If it is a question of which nation has *done* most, no one can deny that the English are that nation," Engels admitted.[58] Industry and trade had made her the world's richest nation. Between 1750 and 1850, the value of goods and services produced in Britain every year—her gross domestic product—had quadrupled, growing more in a hundred years than in the previous thousand.[59] The *Manifesto* emphasized the unprecedented explosion of productive power that Engels and Marx believed would determine political power in the modern world:

> The bourgeoisie, during its rule of scarce one hundred years, has created more massive and more colossal productive forces than have all preceding generations together ... It has been the first to show what man's activity can bring about. It has accomplished wonders far surpassing Egyptian pyramids, Roman aqueducts, and Gothic cathedrals; it has conducted expeditions that put in the shade all former Exoduses of nations and crusades.[60]

Marx and Engels had no doubt that England's capacity to produce would continue to grow by multiples. But they were convinced that the distributive mechanism was fatally flawed and would cause the whole system to collapse. Despite the extraordinary accession of wealth, the abysmally low living standards of the three-fourths of the British people who belonged to the laboring classes had improved only a little. Recent estimates by Gregory Clark and other economic historians suggest that the average wage rose by about one-third between 1750 and 1850 from an extremely low level.[61] True, the laboring classes were now far more numerous, the English population having trebled. And they were not as miserable as their German or French counterparts.

But advances in some areas were balanced by retrogressions elsewhere. For one thing, most of the gain in pay occurred after 1820, and the lion's share went to skilled craftsmen and factory operatives. Any improvement in the wages of unskilled laborers, including farmworkers, was marginal and was offset, as Malthus had feared, by bigger families. Employment was less secure because manufacturing and construction were subject to booms and busts. Hours were longer, and wives and children were more liable to work as well.

Living standards of urban workers were further undermined by the degradation of the physical environment. The mass migration from the country to the city was taking place before the germ theory of disease had been discovered and before garbage collection, sewers, and clean water supplies were commonplace. Despite the greater poverty of rural England, life expectancy in the countryside was about forty-five versus thirty-one or thirty-two in Manchester or Liverpool. Filth and malnutrition simply weren't as deadly in less-contagious circumstances. At a time when cities like Liverpool were expanding at rates between 31 and 47 percent every decade, epidemics posed a constant threat. The richest of the rich were not immune—Prince Albert, Queen Victoria's husband, was carried off by typhoid—but the risks were magnified by poor nutrition and crowding. As the influx of migrants into cities accelerated in the first half of the nineteenth century, the health of the average worker stopped improving with income or actually deteriorated. Life expectancy at birth rose from thirty-five to forty between 1781 and 1851, but raw death rates stopped falling in the 1820s. Infant mortality rose in many urban parishes, and adult height—a measure of childhood nutrition, which is affected by disease as well as diet—of men born in the 1830s and 1840s fell.[62]

Reactionaries and radicals alike wondered if England was suffering from a Midas curse. "This successful industry of England, with its plethoric wealth, has as yet made nobody rich; it is an enchanted wealth," thundered Carlyle.[63] The economic historian Arnold Toynbee argued that the first half of the nineteenth century was "a period as disastrous and as terrible as any through which a nation has ever passed. It was disastrous and terrible, because side by side with a great increase of wealth was seen an enormous increase in pauperism; and production on a vast scale, the result of free competition, led to a rapid alienation of classes and the degradation of a large body of producers."[64]

True, as England's leading philosopher, John Stuart Mill, pointed out, the gradual removal of laws, levies, and licenses that tied the "lower orders" to particular villages, occupations, and masters had increased social mobility: "Human beings are no longer born to their place in life . . . but are free to employ their faculties and such favorable chances as offer, to achieve the lot which may appear to them most desirable."[65] But even Mill, a libertarian with strong Socialist sympathies, could see little improvement in the well-being of most Englishmen: "Hitherto it is questionable if all the mechanical inventions yet made have lightened the day's toil of any human being."[66]

Thus, in the second year of the potato famine in Ireland, the authors of *The Communist Manifesto* repeated Engels's earlier claim that as the nation grew in wealth and power, the condition of its people only worsened: "The modern labourer . . . instead of rising with the progress of industry, sinks deeper and deeper below the conditions of existence of his own class. He becomes a pauper, and pauperism develops more rapidly than population and wealth. And here it becomes evident that the bourgeoisie is unfit any longer to be the ruling class in society. . . . The proletarians have nothing to lose but their chains. They have the world to win. WORKING MEN OF ALL COUNTRIES, UNITE![67]

Having been ejected from France for publishing a satirical sketch of the Prussian king, Marx, his growing family, and the family retainer had been living in Belgium on a publisher's advance for his economics treatise. At the end of his month-long stay in London, Marx had returned to his suburban villa in Brussels, where he promptly put off the task of writing the final version and threw himself into a lecture series . . . on the economics of exploitation. In January, after league officials threatened to hand the assignment to someone else, he finally picked up his pen. Just before news of fighting in Paris between Republicans and the municipal guard reached Great Windmill Street, his partially finished final draft arrived in the mail.

On February 21, the league had one thousand copies of the *Manifesto*, written in German, printed and delivered to the German border with France. All but one copy was promptly confiscated by the Prussian authorities.

Marx and Engels waited impatiently for Armageddon. Like many nineteenth-century romantics, they "saw themselves as living in a general atmosphere of crisis and impending catastrophe" in which *anything* could happen.[68] John of Patmos, the author of the book of Revelation, had supplied them with the perfect finale for modern society and their *Manifesto:* society splits into two diametrically opposed camps, there is a final battle, Rome falls, the oppressed receive justice, the oppressors are judged, and the end of history comes.

History did not end in 1848. The French revolution of that year led not to Socialism or even universal male suffrage, but to the reign of Napoléon III. The declaration of the French Republic resulted in Marx's summary ejection from Belgium and, a few weeks after he had found a new bolt-hole in Paris, persecution by the French authorities. When the Paris police threatened to banish him to a swampy, disease-ridden village hundreds of miles from the capital, Marx objected on grounds of health and began to look around for a country that would take him. In August 1849 he moved to London, that "Patmos of foreign fugitives" and home of the former French king Louis Philippe and countless other political exiles.[69] It would be for only a short time, he consoled himself.

Marx's arrival in London coincided with one of the worst cholera epidemics in the city's history. By the time it had run its course, 14,500 adults and children had died.[70] The outbreak encouraged Henry Mayhew, the journalist, to undertake a remarkable series of newspaper stories about London's poor.[71] A scientist manqué who had a terrible relationship with his father, Mayhew was plump, energetic, and engaging, but absolutely hopeless about money. At thirty-seven, the former actor and cofounder of the humor magazine *Punch* was still recovering from a humiliating bankruptcy that had cost him his London town house and nearly landed him in jail. After months of grinding out pulp fiction with self-mocking titles

such as *The Good Genius That Turned Everything into Gold,* Mayhew saw a
chance for a comeback.

Mayhew's eighty-eight-part series took *Chronicle* readers on a house-
by-house tour in the "very capital of cholera."[72] Jacob's Island was a particularly noxious corner of Bermondsey on the south side of the Thames
immortalized by Dickens in *Oliver Twist.* Mayhew promised readers a sensational portrait of the district's inhabitants "according as they will work,
they can't work, and they won't work."[73] He assured the audience that he
was no "Chartist, Protectionist, Socialist, Communist," which was perfectly
true, but a "mere collector of facts."[74] With a team of assistants and a few
cabmen more or less on retainer, he plunged into the houses with "crazy
wooden galleries . . . with holes from which to look upon the slime beneath; windows, broken and patched, with poles thrust out, on which to
dry the linen that is never there; rooms so small, so filthy, so confined, that
the air would seem to be too tainted even for the dirt and squalor which
they shelter."[75]

Mayhew found that London's working population was by no means
a single monolithic class but a mosaic of distinct and highly specialized
groups.[76] He ignored the city's single biggest occupation—150,000 domestic servants—whose numbers demonstrated how large the rich loomed
in the city's economy. Nor did he take an interest in the 80,000 or so construction workers employed in building railroads, bridges, roads, sewers,
and so on. Instead Mayhew concentrated on a handful of manufacturing
trades. As the historian Gareth Stedman Jones explains, London's labor
market was a marriage of extremes. On the one hand, the city attracted
highly skilled artisans who catered to the wealthy and who earned one-
fourth to one-third more than in other towns, as much as the clerks and
shopkeepers who comprised the "lower" middle class. On the other hand,
it thrived on an uninterrupted influx of unskilled labor. Laborers also
earned higher wages than their counterparts in the provinces, but their living conditions were apt to be worse because of the overcrowded, decrepit
housing in areas like Whitechapel, Stepney, Poplar, Bethnal Green, and
Southwark, which had been exhaustively documented by parliamentary
commissions of the 1840s. Clerks, salespeople, and other white-collar

workers could afford the new omnibuses or trains and were escaping to the fast-growing suburbs. Unskilled workers had no choice but to stay within walking distance of their places of employment.

Competition from provincial towns and other countries was a constant source of pressure to find ways to save on labor costs. The system of "sweating" or piecework, often performed in the worker's own lodging, was tailor-made to keep industries such as dressmaking, tailoring, and shoe manufacturing that would otherwise have migrated out of London on account of its high rents, overheads, and wages. Thus, Stedman Jones concludes, London's poverty, with its sweatshops, overcrowding, chronic unemployment, and reliance on charity, was, in fact, a by-product of London's wealth. The city's rapid growth led to rising land prices, high overheads, and high wages. High wages attracted more waves of unskilled newcomers but also created constant pressure on employers to find ways to replace more expensive labor with cheap labor.

London's needlewomen epitomized the phenomenon, and they were the subjects of Mayhew's most sensational stories. "Never in all history was such a sight seen, or such tales heard," he promised.[77] Using census figures, Mayhew calculated that there were 35,000 needlewomen in London, 21,000 of whom worked in "respectable" dressmaking establishments that ranged from the bespoke to those that catered to the lower middle class. The other 14,000, he wrote, worked in the "dishonorable" or sweated sector.[78] Mayhew contended that piecework rates "of the needlewomen generally are so far below subsistence point, that, in order to support life, it is almost a physical necessity that they must either steal, pawn, or prostitute themselves."[79]

On this occasion, Mayhew was more impresario than observer. In November, with the help of a minister, he organized "a meeting of needlewomen forced to take to the streets." He promised strict privacy of the assembly. Men were barred. Two stenographers took verbatim notes. Under dimmed lights, twenty-five women were given tickets of admission. They mounted the stage and were encouraged to share their sorrows and sufferings. The minister exhorted them to speak freely. To Mayhew's amazement, they did:

The story which follows is perhaps one of the most tragic and touching romances ever read. I must confess that to myself the mental and bodily agony of the poor Magdalene who related it was quite overpowering. She was a tall, fine-grown girl, with remarkably regular features. She told her tale with her face hidden in her hands, and sobbing so loud that it was with difficulty I could catch her words. As she held her hands before her eyes I could see the tears oozing between her fingers. Indeed I never remember to have witnessed such intense grief.[80]

Mayhew's account in the *Morning Chronicle* confirmed Thomas Carlyle's worst fears about modern industrial society, inspiring a choleric rant against economists:

Supply-and-demand, Leave-it-alone, Voluntary Principle, Time will mend it; till British industrial existence seems fast becoming one huge poison-swamp of reeking pestilence physical and moral; a hideous *living* Golgotha of souls and bodies buried alive; such a Curtius' gulf, communicating with the Nether Deeps, as the Sun never saw till now. These scenes, which the Morning Chronicle is bringing home to all minds of men, thanks to it for a service such as Newspapers have seldom done— ought to excite unspeakable reflections in every mind.[81]

Among these unspeakable reflections was the image of a volcano on the verge of eruption. "Do you devour those marvelous revelations of the inferno of misery, of wretchedness, that is smoldering under our feet?" Douglas Jerrold, then editor of *Punch* and Mayhew's father-in-law, asked a friend. "To read of the sufferings of one class, and the avarice, the tyranny, the pocket cannibalism of the others, makes one almost wonder that the world should go on."[82]

Mayhew's series in the *Morning Chronicle*, "Labour and the Poor," ran for the entire year of 1850. When about half of the articles had run, he revealed his larger ultimate aim. He wanted to invent, he confessed, "a new Political Economy, one that will take some little notice of the claims of labour." He justified his ambition by suggesting that an economics that did "justice as well to the workman as to the employer, stands foremost among the desiderata, or the things wanted, in the present age."[83]

Carlyle's friend John Stuart Mill had given precisely the same reason for embarking on his *Principles of Political Economy,* published in 1848, only two years earlier, and already the most-read tract on economics since Adam Smith's *The Wealth of Nations.*

"Claims of Labor have become the question of the day," Mill wrote during the Irish potato famine in 1845, when he conceived the idea for the book.[84] At the time, the thirty-nine-year-old Mill had long been in love with Harriet Taylor, an unhappily married intellectual whom Carlyle described as "pale . . . and passionate and sad-looking" and a "living Romance heroine."[85] As Mill's frustration over Harriet's husband's refusal to grant her a divorce grew, so did his sympathy with her Socialist ideals.

In taking up political economy, Mill hoped to overcome Carlyle's objection that the discipline was "dreary, stolid, dismal, without hope for this world or the next"[86] and Taylor's that it was biased against the working classes. Agreeing with Dickens, Mill saw a particular need to "avoid the hard, abstract mode of treating such questions which has brought discredit upon political economists." He blamed them for enabling "those who are in the wrong to claim, & generally to receive, exclusive credit for high & benevolent feeling."[87]

Mill no doubt had in mind David Ricardo, the brilliant Jewish stockbroker and politician who took up economics as a third career at age thirty-seven. Between 1809 and his untimely death in 1823, Ricardo not only recast the brilliant but often loosely expressed ideas of Adam Smith as an internally consistent, precisely defined set of mathematical principles but also proposed a remarkable number of original ideas concerning the benefits of trade for poor as well as rich nations and the fact that countries prosper most when they specialize. Nonetheless, many potential readers of his *On the Principles of Political Economy and Taxation* were as repelled by Ricardo's tendency to convey his ideas in abstract terms as by his dour conclusions. His iron law of wages—stating that wages may go up or down based on short-run fluctuations in supply and demand but always tend toward subsistence—incorporated Malthus's law of population and ruled out any meaningful gains in real wages.[88]

Mill noted that Ricardo, Smith, and Malthus were all vocal champions of individual political and economic rights, opponents of slavery, and foes of protectionism, monopolies, and landowner privileges. He himself favored unions, universal suffrage, and women's property rights. In response to the economic crisis and social strife of the Hungry Forties, he advocated the repeal of the 50 percent tax on imported grain. The typical laborer spent at least one-third of his meager pay on feeding himself and his family. Mill correctly predicted that once the tax on imports was abolished food prices would decline and real wages would rise. Yet even he remained profoundly pessimistic about the scope of improvements in the lives of workers. Like Carlyle, he was convinced that the repeal of the Corn Laws would only buy time, as the invention of the railroad, the opening up of the North American continent, and the discovery of gold in California had. Such developments, while beneficial, could not repeal the immutable laws by which the world was governed.

Malthus's law of population and Ricardo's iron law of wages and law of diminishing returns—the notion that using more and more labor to farm an acre would produce less and less extra output—all dictated that population would outrun resources and that the nation's wealth could be enlarged only at the expense of the poor, who were doomed to spend "the great gifts of science as rapidly as ... [they] got them in a mere insensate multiplication of the common life." [89] Government could do no more than create conditions in which enlightened self-interest and laws of supply and demand could work efficiently.

For Mill, economies are governed by natural laws, which couldn't be changed by human will, any more than laws of gravity can. "Happily," Mill wrote as he was finishing *Principles* in 1848, "there is nothing in the laws of Value which remains for the present or any future writer to clear up; the theory of the subject is complete." [90]

Henry Mayhew, for one, refused to accept this conclusion. By his lights, Mill had failed in his attempt to turn political economy into a "gay science," that is, a science capable of increasing the sum of human happiness, freedom, or control over circumstances.[91] The fact that Mill had not jettisoned the iron law of wages was all the more reason for trying again. Ultimately, Mayhew did not succeed in mounting a challenge to the classical wage doctrine, and neither did anyone else of his generation. Still, his landmark series on London labor became the unofficial Baedeker for a younger generation of "social investigators" who were inspired by his reporting and shared his desire to learn how much improvement was possible without overturning the social order.

In August 1849, less than two years after Karl Marx had arrived in London amid a cholera epidemic, the whole world seemed to be descending upon his sanctuary to see the Great Exhibition. The first world's fair was the brainchild of another German émigré, Queen Victoria's husband, Prince Albert, but Marx, who was by then living with his wife, Jenny, their three young children, and their housekeeper in two dingy rooms over a shop in Soho, wanted nothing to do with it. He fled to seat G7 in the high-domed reading room of the British Museum with its cathedral-like gloom and refreshing quiet. Ignoring breathless newspaper accounts about the construction of the Crystal Palace in Hyde Park, Marx filled notebook after notebook with quotations, formulas, and disparaging comments as he pored over the works of the English economists Malthus, Ricardo, and James Mill, the father of John Stuart Mill. Let the philistines pray in the bourgeois Pantheon, he told himself. He would have no truck with false idols.

In May 1851, Karl Marx was no longer the dreamy young university student who spent days holed up in his dressing gown writing sonnets to a baron's daughter, or the louche journalist who drank all night in Paris cafés. In the ten years since he had obtained his mail-order doctorate from the University of Jena, he had squandered a surprise inheritance of 6,000 francs from a distant relative. He had started three radical journals, two of which had folded after a single issue. He had never held a job for more than a few months. While his erstwhile protégé, Engels, had produced a best seller, his own magnum opus remained unwritten. He had published, but mostly long-winded polemics against other Socialists. At thirty-two, he was just another unemployed émigré, the head of a large and growing family, forced to beg and borrow from friends. Luckily for him, his guardian angel, Engels, had promised to pursue a career at his family's firm expressly so that Marx could focus on his book full-time.

Meanwhile, as heads of state and other dignitaries swooped into town, Scotland Yard was keeping a close eye on radicals. Judging by a report from a Prussian government spy, the main threat posed by Marx was to Mrs. Beeton's standards of housekeeping:

> Marx lives in one of the worst, therefore one of the cheapest quarters of London. He occupies two rooms. The one looking out on the streets is the salon, and the bedroom is at the back. In the whole apartment there is not one clean and solid piece of furniture. Everything is broken, tattered and torn, with a half inch of dust over everything and the greatest disorder everywhere. In the middle of the salon there is a large old fashioned table covered with an oil cloth, and on it lie manuscripts, books and newspapers as well as the children's toys, the rags and tatters of his wife's sewing basket, several cups with broken ribs, knives, forks, lamps, an inkpot, tumblers, Dutch clay pipes, tobacco ash—in a word everything is topsy-turvy and all on the same table. A seller of second hand goods would be ashamed to give away such a remarkable collection of odds and ends.[92]

The Exhibition season represented a new nadir in Marx's affairs. Though he adored his wife, he had carelessly gotten Helen Demuth, her personal maid and the family housekeeper, pregnant. Jenny, who was pregnant as well, was beside herself. Three months after she gave birth to a sickly girl, the family's housekeeper delivered a bouncing baby boy. To quash the "unspeakable infamies" about the affair already circulating around gossipy émigré circles, Marx had his newborn son whisked off to foster parents in the East End, never to see him again. "The tactlessness of some individuals in this respect is colossal," he complained to a friend.[93] The boy's mother stayed behind to care for the Marx family as before. With home more unbearable than ever, Marx hurried to seat number G7 every morning and stayed until closing.

By the time the Great Exhibition opened on May Day of 1851, Marx
had already begun to doubt that the modern Rome would be overthrown
by her own subjects. Instead of Chartists storming Buckingham Palace,
four million British citizens and tens of thousands of foreigners invaded
Hyde Park to attend the first world's fair. The human wave helped launch
Thomas Cook in the tour business and brought people of all backgrounds
together. "Never before in England had there been so free and general a
mixture of classes as under that roof," crowed one of the many accounts
of the fair published at the time.[94] For Marx, the fair resembled the games
Roman rulers staged to keep the mob entranced. "England seems to be
the rock which breaks the revolutionary waves," he had written in an
earlier column for the *Neue Rheinische Zeitung*. "Every social upheaval in
France . . . is bound to be thwarted by the English bourgeoisie, by Great
Britain's industrial and commercial domination of the world."[95] The Exhibition was meant to encourage commercial competition, which Prince
Albert and some of its other sponsors hoped would foster peace. Marx had
prayed for war: "Only a world war can break old England . . . and bring
the proletariat to power."[96] The worse things got, he reasoned, the better
the odds of revolution.

Still, he was not willing to totally discount the possibility that "the
great advance in production since 1848" might lead to a new and more
deadly crisis. Dismissing the Exhibition as "commodity fetishism," he predicted the "imminent" collapse of the bourgeois order.[97] As he and Engels
had written in their *Manifesto:* "What the bourgeoisie therefore produces,
above all, are its own grave-diggers."[98]

Racing against time so as not to be overtaken by the "inevitable"
revolution—if not in England, then on the Continent—Marx began working furiously on his own book of Revelation, a critique of "what Englishmen call 'The Principles of Political Economy.'"[99] Marx spent most days
scouring the reading room at the British Museum for material for his great
work. To the contemporary questions "How much improvement in living
standards was possible under the modern system of private property and
competition?" and "Could it endure?" Marx *knew* the answers had to be
negative. His challenge now was to prove it.

When he took up economics in 1844, Marx did not set out to show that life under capitalism was awful. A decade of exposés, parliamentary commissions, and Socialist tracts, including Engels's, had already accomplished that. The last thing Marx wanted was to condemn capitalism on moral (that is to say Christian) grounds, as utopian Socialists such as Pierre-Joseph Proudhon, who claimed that "private property is theft," had done. Marx had no intention of converting capitalists as his favorite novelist, Dickens, dreamed of doing with his *Christmas Carol*. In any case, he had long repudiated the notion of any God-given morality and insisted that man could make up his own rules.

The point of his great work was to prove "with mathematical certainty" that the system of private property and free competition couldn't work and hence that "the revolution must come." He wished to reveal "the law of motion of modern society." In doing so, he would expose the doctrines of Smith, Malthus, Ricardo, and Mill as a false religion, just as radical German religion scholars had exposed biblical texts as forgeries and fakes. His subtitle, he decided, would be *A Critique of Political Economy*.[100]

Marx's law of motion did not spring Athena-like from his powerful, brooding mind, as his doctor friend Louis Kugelmann supposed when he sent Marx a marble bust of Zeus as a Christmas present. It was Engels, the journalist, who supplied Marx with the rough draft of his economic theory. Marx's real challenge was to show that the theory was logically consistent as well as empirically plausible.

In the *Manifesto*, Marx and Engels had offered two reasons for capitalism's dysfunction. First, the more wealth that was created, the more miserable the masses would become: "In proportion as capital accumulates, the lot of the laborer must grow worse." Second, the more wealth that was created, "the more extensive and more destructive" the financial and commercial crises that broke out periodically would become.[101]

While the *Manifesto* referred to "ever-decreasing wages" and "ever-increasing burden of toil" as matters of historical fact, in *Das Kapital*, Marx argued that the "law of capitalist accumulation" *requires* wages to fall, the length and intensity of the working day to rise, working conditions to deteriorate, the quality of goods consumed by workers to decline, and the average life span of workers to fall. He did not, however, fall back on the second of his arguments about ever-worsening depressions.[102]

In *Das Kapital,* Marx specifically rejected Malthus's law of population, which, as it happens, is also a theory of how the level of wages is determined. In formulating his law, Malthus had assumed that pay was strictly a function of the size of the labor force. More workers meant more competition among them, hence lower wages. Fewer workers meant the opposite. Engels had already identified the primary objection to Malthus in his 1844 "Outlines of a Critique of Political Economy," namely that poverty could afflict any society, including a Socialist one.

Marx's edifice rests on the assumption that all value, including surplus value, is created by the hours worked by labor. "There is not a single atom of its value that does not owe its existence to unpaid labor." In *Das Kapital,* he cites Mill to support his claim:

> Tools and materials, like other things, have originally cost nothing but labour . . . The labour employed in making the tools and materials being added to the labour afterwards employed in working up the materials by aid of the tools, the sum total gives the whole of the labour employed in the production of the completed commodity . . . *To* replace capital, is to replace nothing but the wages of the labour employed.[103]

Mark Blaug, a historian of economic thought, points out that if only labor hours create value, then installing more efficient machinery, reorganizing the sales force, hiring a more effective CEO, or adopting a better marketing strategy—rather than hiring more production workers—necessarily causes profits to fall. In Marx's scheme, therefore, the only way to keep profits from shrinking is to exploit labor by forcing workers to work more hours without compensating them. As Henry Mayhew detailed in his *Morning Chronicle* series, there are many ways of cutting the real wage. It is crucial for Marx's argument, writes Blaug, that trade unions and governments—"organizations of the exploiting class"—can't reverse the process.[104]

A surprising number of scholars deny that Marx ever claimed that wages would decline over time or that they were tethered to some biological minimum. But they are overlooking what Marx said in so many words on numerous occasions. The inability of workers to earn more when they produce more—or more-valuable products—is precisely what made capitalism unfit to survive.

By asserting that labor was the source of all value, Marx claimed that the owner's income—profit, interest, or managerial salary—was unearned. He did not argue that workers did not need capital—factories, machines, tools, proprietary technology, and the like—to produce the product. Rather he argued that the capital the owner made available was nothing more than the product of *past* labor. But the owner of any resource— whether a horse, a house, or cash—could use it herself. Arguing, as Marx does, that waiting until tomorrow to consume what could be consumed today, risking one's resources, or managing and organizing a business have no value and therefore deserve no compensation is the same as saying that output can be produced without saving, waiting, or taking risks. This is a secular version of the old Christian argument against interest.

The trouble is, as Blaug points out, that this is just another way of saying that only labor adds value to output—the very statement that Marx set out to prove in the first place—and not an independent proof.

Marx compiled an impressive array of evidence, from Blue Books, newspapers, the *Economist,* and elsewhere, to show that the living standards of workers were wretched and working conditions horrendous during the second half of the eighteenth and first half of the nineteenth centuries. But he did not succeed in showing either that average wages or living standards were declining in the 1850s and 1860s, when he was writing *Das Kapital,* or, more to the point, that there was some reason for thinking that they would *necessarily* decline.

Had Marx stepped outside and taken a good look around like Henry Mayhew, or engaged brilliant contemporaries such as John Stuart Mill who were grappling with the same questions, he might have seen that the world wasn't working the way he and Engels had predicted. The middle class was

growing, not disappearing. Financial panics and industrial slumps weren't
getting worse.

When the Great Exhibition of 1862 closed, the "great festival" refused
to disband. A businessman bought the Crystal Palace, had it disassembled
and carted to Sydenham in South London, and rebuilt it on an even more
monstrous scale. Much to Marx's disgust, the new Crystal Palace opened
as a kind of Victorian Disney World. Worse, the economy boomed. As
Marx had to admit, "It is as if this period had found Fortunatas' purse."
There had been a "titanic advance of production" even faster in the second
ten years than in the first:

> No period of modern society is so favorable for the study of capitalist
> accumulation as the period of the last 20 years . . . But of all countries
> England again furnishes the classical example, because it holds the fore-
> most place in the world-market, because capitalist production is here
> alone completely developed, and lastly, because the introduction of the
> Free-trade millennium since 1846 has cut off the last retreat of vulgar
> economy.[105]

More fatal to Marx's theory, real wages weren't falling as capital accumulated in the form of factories, buildings, railroads, and bridges. In
contrast to the decades before the 1840s, when increases in real wages were
largely limited to skilled workers, and the effect on living standards was
offset by more unemployment, longer hours, and bigger families, the gains
in the 1850s and 1860s were dramatic, unambiguous, and widely discussed
at the time. The Victorian statistician Robert Giffen referred to the "un-
doubted" nature of the "increase of material prosperity" from the mid-
1840s through the mid-1870s.[106] Robert Dudley Baxter, a solicitor and
statistician, depicted the distribution of income in 1867 with an extinct
volcano that rose twelve thousand feet above sea level, "with its long low
base of laboring population, with its uplands of the middle classes, and
with the towering peaks and summits of those with princely incomes."[107]
The Peak of Tenerife struck Baxter as a perfect metaphor for describing

who got what. Still, his data show that by 1867, labor's share of national income was rising.

Scholars have since corroborated these contemporary observations. As early as 1963, Eric Hobsbawm, the Marxist economic historian, admitted that "the debate is entirely about what happened in the period which *ended* by common consent sometime between 1842 and 1845."[108] More recently, Charles Feinstein, an economic historian on the "pessimist" side of a long-running debate on the effects of the industrial revolution, concluded that real wages "at last started an ascent to a new height" in the 1840s.[109]

Marx never did step outside. He never bothered to learn English well.[110] His world was restricted to a small circle of like-minded émigrés. His contacts with English working-class leaders were superficial. He never exposed his ideas to people who could challenge him on equal terms. His interaction with economists—"commercial travelers for the great firm of Free-trade"[111] as he called them—whose ideas he wished to demolish, was nonexistent. He never met or conducted a scientific correspondence with the geniuses—John Stuart Mill, the philosopher; Charles Darwin, the biologist; Herbert Spencer, the sociologist; George Eliot, the writer; among them—who lived (and debated) a mile or two from him. Astonishingly for the best friend of a factory owner and the author of some of the most impassioned descriptions of mechanization's horrors, Marx never visited a single English factory—or any factory at all until he went on a guided tour of a porcelain manufactory near Carlsbad, where he took the waters toward the end of his life.[112]

At Engels's insistence, in 1859 Marx reluctantly published a preview of his unfinished magnum opus. The thin volume, called *A Contribution to the Critique of Political Economy*, was greeted with surprise, embarrassment, and virtually no reviews except ones that Engels wrote anonymously at Marx's behest.[113]

Marx had frequently justified his decision to remain in England—and even to seek British citizenship—by pointing to the advantages of

London, capital of the modern world, for studying the evolution of society and glimpsing its future. But Isaiah Berlin, himself an émigré, wrote that "he might just as well have spent his exile in Madagascar, provided that a regular supply of books, journals and government reports could have been secured." By 1851, when he started to work seriously on the critique that he boasted would demolish English economics, Marx's ideas and attitudes were "set and hardly changed at all" over the next fifteen or more years.[114]

When Marx took up the idea of "providing a complete account and explanation of the rise and imminent fall of the capitalist system,"[115] his eyesight was so bad that he was forced to hold books and newspapers a few inches from his face. One wonders what effect his myopia had on his ideas. Democritus, the subject of his doctoral dissertation, was said to have blinded himself deliberately. In some versions of his legend, the Greek philosopher is motivated by a desire to avoid being tempted by beautiful women. In others, he wants to shut out the messy, confusing, shifting world of facts so that he can contemplate the images and ideas in his own head without these bothersome distractions.

One might think that his family's climb from renters of rooms over a store to rate-paying owners of a London town house would have made Marx uneasy about his theory. In the twenty years since he had set out to prove that capitalism could not work, Marx himself had evolved from bohemian to bourgeois. He no longer favored the immediate abolition of the rights of inheritance in the Communist program.[116] The Marxes used one of several legacies to trade their "old hole in Soho" for an "attractive house" in one of the new middle-class developments near Hampstead Heath. It was so new that they found there was no paved road, no gas street lights, and no omnibuses; only heaps of rubbish, piles of rock, and mud.

Marx often said that there was something rotten about a system that increased wealth without reducing misery, yet it did not seem to strike him that misery can sometimes increase with wealth. He assumed that London's slums, which were becoming more Dickensian with each passing decade, were proof that the economy couldn't deliver a decent standard of living for ordinary people. On the contrary, explains Gareth Stedman Jones, the housing crisis was an unwelcome by-product of London's helter-skelter growth, growing prosperity, and voracious demand for unskilled labor. The key fact is that the mid-Victorian building frenzy involved an orgy of demolition. Between 1830 and 1870, thousands of acres in central London were cleared, mostly in the poor districts where land was cheap, to expand the London docks, lay railway lines, build New Oxford Street, dig the sewers and water pipes, and, in the 1860s, excavate the first stretches of the London tube. So, just as tens of thousands of migrants were flocking to the city in search of work, the supply of housing within walking distance of London's industrial areas was plummeting. As a result, workers were crowded into ever more dilapidated, ever tighter, ever more expensive quarters. Once the demolition stopped and white-collar workers began to commute from the suburbs by rail, the housing crisis began to ease.

The Exhibition season of 1862 coincided with another low point in Marx's financial affairs. Horace Greeley, the publisher of the *New York Tribune,* had dropped his column, which, though entirely ghostwritten by Engels, had supplied Marx with extra cash. At one point, his money woes became so dire that he applied for a job as a railway clerk, only to be rejected for "bad handwriting" and not speaking English, and briefly considered immigrating to America. Luckily, he was like an oyster that needed a bit of grit to make his pearls. With his mind on money, he was soon writing a long essay on economics and filling up notebooks again, complaining all the while that he felt like "a machine condemned to devour books and then throw them, in a changed form, on the dunghill of history."[117] He also decided on a title for his great work: *Das Kapital.*[118]

The hoopla surrounding the Exhibition continued to depress Marx. He would have sympathized with Fyodor Dostoyevsky's reaction; the Russian novelist called the glass palace "a Biblical sight, something to do with Babylon, some prophecy out of the Apocalypse being fulfilled before your very eyes."[119] Yet within a year or two, Marx's fortunes turned up again. Thanks to several unexpected legacies as well as a £375 annual subsidy from Engels, he was able to move his family to an even bigger and more imposing town house and was soon spending £500 to £600 a year, something that more than 98 percent of English families could not afford to do.[120]

Marx had almost forgotten about the Day of Judgment when it dawned.

The launch of the eleven-thousand-ton warship the HMS *Northumberland* on April 17, 1866, ought to have been a day of pride, a reminder of Great Britain's industrial and commercial domination of the world. Instead it was a fiasco. The *Northumberland* had been on the slips in the Millwall Iron Works yard for nearly five years. On the day of the launch, her unusually heavy weight caused her to slip off the railing—a portent, people understood later, of the precarious condition of the shipping firms and shipbuilders.

Less than a month later, on Thursday afternoon, May 10, in the first week of the London boating season, a frightful rumor swirled through the city. The Rolls-Royce of merchant banks, Overend, Gurney & Company, considered by the average citizen to be as solid as the Royal Mint, had failed. "It is impossible to describe the terror and anxiety which took possession of men's minds for the remainder of that and the whole of the succeeding day," wrote the London *Times*'s financial correspondent. "No man felt safe." By ten o'clock the following morning, a horde of "struggling and half frantic creditors" of both sexes and seemingly all stations of life invaded the financial district. "At noon the tumult became a rout. The doors of the most respectable Banking Houses were besieged . . . and throngs heaving and tumbling about Lombard Street made that narrow thoroughfare impassable."[121]

The *New York Times* bureau chief dashed off a telegram to his editors to convey that this was "a more fearful panic than has been known in the British metropolis within the memory of man." Before an extra battalion of constables could be called out to control the crowd and before the Chancellor of the Exchequer could authorize the suspension of the Bank Charter Act, the Bank of England had lost 93 percent of its cash reserves, the British money market was frozen solid, and scores of banks and businesses that lived on credit were facing ruin. "Englishmen have been running mad on speculation . . . The day of reckoning has arrived and blank panic and blue dismay sit on the faces of all our bankers, capitalists and merchants." [122]

Among the first victims of the panic were the owners of the Millwall shipyard. The boom in shipbuilding, fueled by a worldwide arms race and trade, had more than doubled employment in London shipyards between 1861 and 1865. [123] "The magnates of this trade had not only over-produced beyond all measure during the overtrading time, but they had, besides, engaged in enormous contracts on the speculation that credit would be forthcoming," Marx gloated. [124]

By the time of the Overend collapse, new orders were drying up. In fact, Overend may have been pushed over the edge because "they covered the seas with their ships" and "were incurring huge losses on their fleet of steamships." Other casualties included the legendary railway contractors Peto and Betts. True, the most immediate victims of the panic were gullible investors and "countless swindling companies" that had sprung up to take advantage of cheap money. But the crisis of confidence forced the Bank of England to raise its benchmark interest rate from 6 percent to a crushing 10 percent, "the classic panic rate," [125] which persisted through the summer. A play called *One Hundred Thousand Pounds* closed after a brief run. The *Times* didn't even bother to review it. The boom was over.

When news of Black Friday reached Marx via his afternoon paper, he was in his study in North London pondering a financial crisis closer to home. One Modena Villas, where he and his family had recently moved, was a pretentious affair of the kind sprouting up all over London's periphery, far too pricey for an unemployed journalist who had long since stopped accepting assignments in order to finish his book. Marx had rationalized the extravagance as necessary for his teenage daughters "to establish themselves socially." Now, alas, he was broke again and his rent was overdue. So, unfortunately, was *Das Kapital*.

For nearly fifteen years, Marx had been assuring his best friend and patron that his grandiose "Critique of Political Economy" was "virtually finished, that he was ready to "reveal the law of motion of modern society," that he would drive a stake through the heart of English "political economy." Now Engels, who had kept his nose to the grindstone in Manchester for fifteen years to support him, was becoming restive.

In truth, the glitter of England's prosperity had cast a pall on Marx's project. He had written very little since 1863. A series of windfalls had purchased temporary spells of independence, but now he was back on Engels's dole, and, for the first time, the angelic Engels was showing signs of impatience. Marx had been putting him off with graphic descriptions of a series of afflictions worthy of Job: rheumatism, liver trouble, influenza, toothache, impudent creditors, an outbreak of boils of truly biblical proportions—the list went on and on. In April 1866, Marx confessed, "Being unwell I am unable to write." On the day after Christmas, he complained of "not writing at all for so long." Around Easter, writing from the seaside in Margate, he admitted to having "lived for my health's sake alone" for "more than a month." [126]

Engels suspected, accurately as it turns out, that the real source of Marx's troubles was "dragging that damned book around" for too long: "I hope you are happily over your rheumatism and faceache and are once more *sitting diligently* over the book," he wrote on May 1. "How is it coming on and when will the first volume be ready?" [127] Since *Das Kapital* was *not* coming on, Marx retreated into a sulky silence.

Like a shot of adrenaline, Black Friday had a galvanizing effect that no amount of nagging by Engels had ever achieved. Within days, the prophet was back at his desk writing furiously. In early July, he was able to report to Engels, "I have had my nose properly to the grindstone again over the past two weeks," and to predict that he would be able to deliver the tardy manuscript "by the end of August." [128]

Who can blame the author of an apocalyptic text holding back until the time was right? By the time Marx was composing it, his melodramatic prophecy, "The death knell of capitalist private property sounds. The expropriators will be expropriated," sounded almost plausible. Yet when he composed his famous penultimate chapter on "The General Law of Capitalist Accumulation," he felt forced to fudge in order to make his case that the poor had gotten poorer. Quoting Gladstone on the "astonishing" and "incredible" surge in taxable income between 1853 and 1863, Marx has the liberal prime minister referring to "this intoxicating augmentation of wealth and power . . . entirely confined to classes of property." [129] The text of the speech, printed in the *Times of London*, shows that Gladstone actually said the opposite:

"I should look with some degree of pain, and with much apprehension, upon this extraordinary and almost intoxicating growth, if it were my belief that it is confined to the class of persons who may be described as in easy circumstances," he said, adding that, thanks to the rapid growth of untaxed income, "the average condition of the British laborer, we have the happiness to know, has improved during the last 20 years in a degree which we know to be extraordinary, and which we may almost pronounce to be unexampled in the history of any country and of any age." [130]

Marx's prediction that his manuscript would be finished by the end of the summer proved wildly optimistic, but fifteen months after Black Friday, in August 1867, he was able to report to Engels that he had put the final set of galleys in the mail to the German publisher. In his note, he alluded in passing to a famous short story by the French novelist Honoré de Balzac. An artist believes a painting to be a masterpiece because he has been perfecting it for years. After unveiling the painting he looks at it for a moment before staggering back. "'Nothing! Nothing! After ten years of work.' He sat down and wept." [131] Alas, as Marx feared, "The Unknown Masterpiece" was an apt metaphor for his economic theory. His "mathematical proof" was greeted by an eerie silence. And in the worst economic crisis of the modern age, the great twentieth-century economist John Maynard Keynes would dismiss *Das Kapital* as "an obsolete economic textbook which I know to be not only scientifically erroneous but without interest or application to the modern world." [132]

Chapter II

Must There Be a Proletariat?
Marshall's Patron Saint

The horseman serves the horse,
The neat-herd serves the neat,
The merchant serves the purse,
The eater serves his meat;
'Tis the day of the chattel,
Web to weave, and corn to grind;
Things are in the saddle,
And ride mankind.

—Ralph Waldo Emerson,
from "Ode, Inscribed to William H. Channing"[1]

The desire to put mankind into the saddle is the mainspring of most economic study.

—Alfred Marshall[2]

During the severe winter of 1866–1867, as many as a thousand men congregated daily at one of several buildings in London's East End. When the doors parted, the crowd surged forward, shoving and shouting, to fight for tickets. From the frenzied assault and the bitter expressions of those who were unsuccessful, a passerby might have assumed that a boxing match or dogfight was starting. But there was no brightly lit ring inside, only the muddy courtyard of a parish workhouse. The yard was divided into pens furnished with large paving stones. A ticket entitled the bearer to sit on one of these slabs, seize a heavy hammer, and break up the grime-encrusted granite. Five bushels of macadam earned him three pennies and a loaf of bread.[3]

The men who besieged the workhouses that January were not typical of the sickly, ragged clientele ordinarily associated with these despised institutions. They were sturdy fellows in good coats. Until a few months earlier, they had been earning a pound or two a week in the shipyards or railway tunnels and highways—more than enough to house a family of five, eat plenty of beef and butter, drink beer, and even accumulate a tiny nest egg.[4] That was before Black Friday brought building on land and sea and underground to an eerie standstill and an avalanche of bankruptcies deprived thousands of their jobs; before a cholera epidemic, a freak freeze that shut down the docks for weeks, and a doubling of bread prices; before the savings of a lifetime were drained away, the last of the household objects pawned, and help from relatives exhausted.

The poorest parishes were turning away hundreds every day while hard-pressed taxpayers like Karl Marx worried that the rising poor rates would ruin them too. Despite an outpouring of donations, private charities were overwhelmed. "What that distress is no one knows," wrote Florence Nightingale, the heiress and hospital reformer, to a friend in January 1867:

> It is not only that there are 20,000 people out of employment at the East End, as it is paraded in every newspaper. It is that, in every parish, not less than twice and sometimes five times the usual number are on the Poor Law books. It *is* that all the workhouses are hospitals. It *is* that the ragged schools instead of being able to give one meal a day are in danger of being shut up. And this all over Marylebone, St. Pancras, the Strand, and the South of London.[5]

Bread riots broke out in Greenwich, and bakers and other small shopkeepers threatened to arm themselves against angry mobs.[6] In May, thousands of East End residents battled mounted police in Hyde Park, ostensibly to show their support for the Second Reform Act and the workman's right to vote, but mostly to vent their frustration and fury at the rich.[7]

Middle-class Londoners could hardly avoid knowing of the distress in their midst, for they were living in the new information age, bombarded by mail deliveries five times a day, newspapers, books, journals, lectures, and sermons. A new generation of reporters inspired by the examples of Henry Mayhew, Charles Dickens, and other journalists of the 1840s filled the pages of the *Daily News,* the *Morning Star,* the *Pall Mall Gazette,* the *Westminster Review, Household Words,* the Tory *Daily Mail,* and the liberal *Times* with sensational eyewitness accounts and firsthand investigations in the East End. Reporters disguised themselves as down-and-out workmen and spent nights in the poorhouse in order to describe its horrors. Robert Giffen, editor at the liberal *Daily News,* was becoming one of the foremost statisticians of his day. His first major academic article had celebrated the tripling of national wealth between 1845 and 1865, but his second, written in 1867, was markedly different in tone and point of view, an attack on harshly regressive tax proposals that fell on the "necessaries of the poor." What upset Giffen about the 1866–67 depression, writes his biographer Roger Mason, is that its chief victims had mostly worked, had saved, and had obeyed the law while the more fortunate had donated generously to charity. But virtue had not sufficed to prevent widespread misery.[8]

The resurgence of hunger, homelessness, and disease in the midst of great wealth radicalized the generation that had grown up during the boom and had taken affluence and progress for granted. Playwrights wrote dramas with proletarian heroes. Poets published works of social criticism. Professors and ministers used their pulpits to denounce British society. Typical of such jeremiads was that of the blind Liberal reformer Henry Fawcett, who held a chair in political economy at the University of Cambridge:

> We are told that our exports and imports are rapidly increasing; glowing descriptions are given of an Empire upon which the sun never sets, and of a commerce which extends over the world. Our mercantile marine

is ever increasing; manufactories are augmenting in number and in magnitude. All the evidences of growing luxury are around us; there are more splendid equipages in the parks and the style of living is each year becoming more sumptuous ... But let us look on another side of the picture; and what do we then observe? Side by side with this vast wealth, closely contiguous to all this sinful luxury there stalks the fearful specter of widespread poverty, and of growing pauperism! Visit the greatest centres of commerce and trade, and what will be observed? The direst poverty always accompanying the greatest wealth![9]

Filled with Christian guilt and the desire to do good, university graduates who had earlier anticipated becoming missionaries in remote corners of the empire were discovering that a great deal of good needed to be done at home. William Henry Fremantle, the author of *The World as the Subject of Redemption,* became the vicar in one of London's poorest parishes, St. Mary's, that year. A walk through the East End during the cholera epidemic convinced Thomas Barnardo, a member of an evangelical sect, to build orphanages for pauper children instead of going to China to convert the Chinese. A similar experience inspired "General" William Booth, the author of *In Darkest England and the Way Out,* to organize a Salvation Army. Samuel Barnett, an Oxford scholar, founded the University Settlers Association to encourage university students to live among the poor running soup kitchens and evening classes.

Missionaries in their own land, these young men and women strove to be scientific rather than sentimental. Their vocation was not dispensing charity but converting the poor to middle-class values and habits. As Edward Denison, an Oxford graduate, remarked in 1867: "By giving alms you keep them permanently crooked. Build school-houses, pay teachers, give prizes, frame workmen's clubs; help them to help themselves."[10]

A young man with delicate features, silky blond hair, and shining blue eyes boarded the Glasgow-bound Great Northern Railway at London's Euston Station. It was early June 1867. He was carrying only a walking stick and a rucksack crammed with books. His fellow passengers might have taken him for a curate or schoolmaster on a mountaineering holiday. But when the train reached Manchester, the young man put his rucksack on, jumped down onto the platform, and disappeared in the crowd.

Before resuming his journey north to the Scottish highlands, Alfred Marshall, a twenty-four-year-old mathematician and fellow of St. John's College in Cambridge, spent hours walking through factory districts and the surrounding slums "looking into the faces of the poorest people." He was debating whether to make German philosophy or Austrian psychology his life's work. These were his first steps away from metaphysics and the beginning of a dogged pursuit of social reality. He later said that these walks forced him to consider the "justification of existing conditions of society."[11]

In Manchester, Marshall found the smoky brown sky, muddy brown streets, and long piles of warehouses, cavernous mills, and insalubrious tenements—all within a few hundred yards of glittering shops, gracious parks, and grand hotels—that novels such as Elizabeth Gaskell's *North and South* had led him to expect. In the narrow backstreets he encountered sallow, undersized men and stunted, pale factory girls with thin shawls and hair flecked with wisps of cotton. The sight of "so much want" amid "so much wealth" prompted Marshall to ask whether the existence of a proletariat was indeed "a necessity of nature," as he had been taught to believe. "Why not make every man a gentleman?" he asked himself.[12]

Marshall, who lacked the plummy accent and easy manners of other fellows of St. John's College, sometimes compared his discovery of poverty to that of original sin and his ultimate embrace of economics to a religious conversion. But although poverty first occurred to him as a subject of study after the panic of 1866, the implication that he had had to wait until then to look into the faces of poor people was grossly misleading.[13] His maternal grandfather was a butcher and his paternal grandfather a bankrupt. His father and uncles started life as penniless orphans. William Marshall had put down "gentleman" as his occupation on his marriage license, but he had never risen above the modest position of cashier at the Bank of England. His son Alfred was born not, as he later intimated, in an upscale suburb but in Bermondsey, one of London's most notorious slums, in the shadow of a tannery. When the Marshalls moved to the lower-middle-class Clapham, they took a house opposite a gasworks.

Thanks to his precocious intelligence and his father's efforts to convince a director of the bank to sponsor his education, Marshall was admitted to Merchant Taylors', a private school in the City that catered to the sons of bankers and stockbrokers. From the age of eight, he commuted daily by omnibus, ferry, and foot through the most noxious manufacturing districts and slums bordering the Thames. Marshall had been looking into the faces of poor people all his life.

In Charles Dickens's *Great Expectations*, published in 1861, the year Marshall graduated from Merchant Taylors', the diminutive orphan hero, Pip, makes what he describes as a "lunatic confession." After swearing his confidante to absolute secrecy three times over, he whispers, "I want to be a gentleman."[14] His playmate Biddy is as nonplussed as if Pip, on the verge of being apprenticed to a blacksmith, had expressed ambitions to become the Pope. Indeed, to make his hero's mad dream come true, Dickens had to invent convicts on a foggy moor, a haughty heiress, a haunted mansion, a mysterious legacy, and a secret benefactor. Even in an age that celebrated the self-made man, the notion that a boy like Pip—never mind the whole mass of Pips—could join the middle class was understood to be the stuff of pure fantasy or eccentric utopian vision, as divorced from real life as Dickens's phantasmagoric novel. As an editorialist for the *Times* observed dryly in 1859, "Ninety nine people in a hundred cannot 'get on' in life but are tied by birth, education or circumstances to a lower position, where they must stay."[15]

Yet there were signs of motion and upheaval. The question of who could become a gentleman, and how, became one of the great recurring themes of Victorian fiction, observes Theodore Huppon. A gentleman was defined by birth and occupation and by a liberal, that is to say non-vocational, education. That excluded anyone who worked with his hands, including skilled artisans, actors, and artists, or engaged in trade (unless on a very grand scale). Miss Marrable in Anthony Trollope's *The Vicar of Bullhampton* "had an idea that the son of a gentleman, if he intended to maintain his rank as a gentleman, should earn his income as a clergyman, or as a barrister, or as a soldier, or as a sailor."[16] The explosion of white-collar professions was blurring the old lines of demarcation. Why else would Miss Marrable have needed to lay down the law? Doctors, architects, journalists, teachers, engineers, and clerks were pushing themselves forward, demanding a right to the label.[17]

A working gentleman's occupation had to allow him enough free time to think of something other than paying the bills, and his income had to suffice to provide his sons with educations and his daughters with gentlemen husbands. Yet exactly what such an amount might be was also a matter of much debate. The paupers in Trollope's *The Warden* are convinced that £100 a year was enough to transform them all into gentlemen, but when the unworldly warden threatens to retire on £160 a year, his practical son-in-law chides him for imagining that he could live decently on such a mere pittance.[18] Alfred Marshall's father supported a wife and four children on £250 per annum,[19] but Karl Marx, admittedly no great manager of money, couldn't keep up middle-class appearances on twice that amount.[20] In 1867 gentlemanly incomes were few and far between. Only one in fourteen British households had incomes of £100 or more.[21]

Yet even Miss Marrable might have agreed that a fellow of a Cambridge college qualified. All fifty-six fellows of St. John's College were entitled to an annual dividend from the college's endowment that rose from about £210 in 1865 to £300 in 1872—as well as rooms and the services of a college servant.[22] A daily living allowance covered dinner at "high table," which usually consisted of two courses, including a joint and vegetables, pies and puddings, followed by a large cheese that traveled down the table on castors. Twice a week a third course of soup or fish was added. Most fellows supplemented their fellowship income with exam coaching fees or specific college jobs such as lecturer or bursar. For a single man with no wife and children—fellows were required to remain celibate—college duties still left many hours for research, writing, and stimulating conversation and an income that permitted regular travel, decent clothes, a personal library, and a few pictures or bibelots—the requisites, in short, of a gentleman's life.

• • •

Alfred Marshall's metamorphosis from a pale, anxious, underfed, badly dressed scholarship boy into a Cambridge don was nearly as remarkable as Pip's transformation from village blacksmith's apprentice into partner in a joint stock company. His father had gone to work in a City brokerage at sixteen. His brother Charles, just fourteen months his senior, was sent to India at seventeen to work for a silk manufacturer. His sister Agnes followed Charles to India, in order to find a husband but died instead.

Like many frustrated Victorian fathers, Marshall's tried to live vicariously through his gifted son. Committed to educating Alfred for the ministry, William Marshall got his employer to foot the tuition at a good preparatory school. He was "cast in the mould of the strictest Evangelicals, bony neck, bristly projecting chin,"[23] a domestic tyrant who bullied his wife and children. A night owl, he often kept Alfred up until eleven, drilling him in Hebrew, Greek, and Latin.[24]

Not surprisingly, the boy suffered from panic attacks and migraines. A classmate remembered that he was "small and pale, badly dressed, and looked overworked." Shy and nearly friendless, Marshall revealed "a genius for mathematics, a subject that his father despised," and acquired a lifelong distaste for classical languages. "Alfred would conceal Potts's Euclid in his pocket as he walked to and from school. He read a proposition and then worked it out in his mind as he walked along."[25]

Merchant Taylors' School was relatively cheap and heavily subsidized, but even with a salary of £250, William Marshall could barely afford the £20 per annum required to cover his son's out-of-pocket expenses as a day student.[26] Yet the senior Marshall was willing to endure—and impose—the strictest economies to send Alfred there, because success at Merchant Taylors' guaranteed a full scholarship to study classics at Oxford, no small prize at a time when a university education was a luxury that only one in five hundred young men of his son's generation could afford. Even more important, under soon to be abolished statutes, the Oxford scholarship came with a virtual guarantee of a lifetime fellowship in classics at one of its colleges or entrée into the church, the civil service, or the faculty of the most prestigious preparatory schools.

When Marshall announced his intention of turning down the Oxford scholarship and studying mathematics at Cambridge instead, his father raged, threatened, and cajoled. Only a substantial loan from an uncle in Australia and a mathematics scholarship enabled Marshall to defy parental authority and pursue his dream. When the seventeen-year-old went up to take his scholarship exam, he walked along the river Cam shouting with joy at his impending liberation.

At the end of three years at St. John's, there was another race to run, namely a grueling sporting event known as the Mathematical Tripos. Leslie Stephen, who was Marshall's contemporary at Cambridge and the future father of Virginia Woolf, estimated that a second-place finish such as Marshall's was worth as much as a £5,000 inheritance—one-half million dollars in today's money—more than enough to get a leg up in life.[27] Marshall's reward was immediate election to a lifetime fellowship at his college, which gave him the right to live at the college and to collect coaching and lecture fees (worth another £2,500 in Stephen's reckoning). After a year of moonlighting at a preparatory school to repay his uncle's loan, Marshall was, for the first time in his life, truly financially independent and free to do as he liked.

How to best use his freedom was the great question. Mathematics was beginning to bore him. As Marshall sat high up in the pure Highland air reading Immanuel Kant ("The only man I ever worshipped"[28]), the world below was hidden in mist. Yet the faces of the poor and images of drudgery and privation continued to haunt him. Like Pip, Alfred Marshall had shot up but could not forget those left behind.

Marshall had returned to Cambridge from Scotland in October 1867, "brown and strong and upright."[29] As an undergraduate he had been excluded from all the social clubs and private gatherings in dons' rooms that constituted the most valuable parts of a Cambridge education. But now that he had achieved intellectual distinction, he was invited to join the Grote Club, a group of university radicals who met regularly to discuss political, scientific, and social questions. Their leader was Henry Sidgwick, a charismatic philosopher four years Marshall's senior who quickly spotted

Marshall's talent and took him under his wing. "I was fashioned by him," Marshall acknowledged. His own father had almost squeezed the life out of him, but Sidgwick "helped me to live." [30]

With Sidgwick as intellectual guide, Marshall plunged into German metaphysics, evolutionary biology, and psychology, rising at five to read every day. He spent some months in Dresden and Berlin, where, according to biographer Peter Groeneweger, he "fell under the spell of Hegel's *Philosophy of History*." [31] Like the young Hegel and Marx, he found Hegel's message that individuals should govern themselves according to their own conscience, not in blind obedience to authority, compelling. He absorbed an evolutionary view of society from Charles Darwin's *On the Origin of Species*, which appeared in 1859, and Herbert Spencer's *Synthetic Philosophy*, published in 1862. An interest in psychology was stimulated by the possibility of "the higher and more rapid development of human faculties." [32] The young man whose chances in life had turned on access to first-rate education was coming to the conclusion that the greatest obstacles to man's mental and moral development were material.

He began to think of himself as a "Socialist." In the 1860s, the term implied an interest in social reform or membership in a communal sect, while the equally expansive label of "Communist" encompassed everyone who thought that things couldn't get better unless the whole system of private property and competition was torn down. [33] When Marshall questioned Sidgwick about overcoming class divisions, his mentor used to gently chide him, "Ah, if you understood political economy you would not say that." Marshall took the hint. "It was my desire to know what was practical in social reform by State and other agencies that led me to read Adam Smith, Mill, Marx and LaSalle," he later recalled. He began his education by reading John Stuart Mill's *Principles of Political Economy*, then in its sixth edition, and "got much excited about it." [34]

His interest was intensified by the unexpected passage of the Reform Act of 1867, which, in a single stroke, turned England into a democracy. The act did more than double the size of the electorate by extending the franchise to some 888,000 adult men, mostly skilled craftsmen and shopkeepers, who paid at least £10 a year in rent or property tax. It brought the working classes into the political system and made democratic government the only acceptable form of government. Though it ignored the 3 million factory operatives, day laborers, and farm workers—and, of course, the entire female sex—twentieth-century historian Gertrude Himmelfarb emphasizes that the Reform Act nonetheless lent the notion of universal suffrage an aura of inevitability.[35] Marshall was troubled, though, by the contrast between the ideal of full citizenship and the reality of material squalor and deprivation that prevented most of his countrymen from taking full advantage of their civic freedoms.

"Shooting up," as Marshall had done, can provoke feelings of guilt or a sense of obligation. Victorian fiction is populated by the "double" who shares the hero's attributes and aspirations but is condemned to stay put while the other shoots up. When the American journalist and writer Henry James explored London on foot in 1869, Hyacinth Robinson, the protagonist of James's 1886 novel about terrorists, seemed to jump "out of the London pavement." James was watching the parade of brilliantly dressed figures, carriages, brilliantly lit mansions and theaters, the clubs and picture galleries emitting agreeable gusts of sound with a sense of doors that "opened into light and warmth and cheer, into good and charming relations," when he conceived a young man very much like himself "watching the same public show . . . I had watched myself," including "all the swarming facts" that spoke of "freedom and ease, knowledge and power, money, opportunity and satiety," with only one difference: the bookbinder turned bomber in *The Princess Casamassima* would "be able to revolve around them but at the most respectful of distances and with every door of approach shut in his face."[36]

Having been admitted to the rarified world of freedom, opportunity, knowledge, and ease, if not power or great wealth, Marshall kept the face of his double where he could see it every day:

> I saw in a shop-window a small oil painting [of a man's face with a
> strikingly gaunt and wistful expression, as of one "down and out"] and
> bought it for a few shillings. I set it up above the chimney piece in my

room in college and thenceforward called it my patron saint, and devoted myself to trying how to fit men like that for heaven.[37]

As Marshall studied the works of the founders of political economy, "economics grew and grew in practical urgency, not so much in relation to the growth of wealth as to the quality of life; and I settled down to it." The "settling" took a while. He found "the dry land of facts" intellectually unappetizing and socially unappealing. When he was asked to take over some lectures on political economy, Marshall agreed reluctantly. "I taught economics . . . but repelled with indignation the suggestion that I was an economist . . . 'I am a philosopher straying in a foreign land.'"[38]

When Marshall began to study economics seriously in 1867, his mentor Sidgwick was convinced that the "halcyon days of Political Economy had passed away."[39] After the success of the 1846 Corn Law repeal, which was followed by a period of low food prices, political economy had a brief turn as "a true science on par with astronomy."[40] But the economic crisis and political upheavals of the 1860s revived the old animus against the discipline among intellectuals. Going a step beyond Carlyle's epithet "the dismal science," John Ruskin, the art historian, dismissed political economics as "that bastard science" and, like Dickens, called for a new economics; "a real science of political economy."[41] The fundamental problem, observed Himmelfarb, was that "the science of riches" clashed with the evangelicalism of the late Victorian era.[42] Victorians were repelled by the notion that greed was good or that the invisible hand of competition guaranteed the best of all possible outcomes for society as a whole.

With the advent of the franchise for working men, both political parties were courting the labor vote. But "political economy" was invoked to oppose every reform—whether higher pay for farm laborers or relief for the poor—on the grounds that it would slow down the growth of the nation's wealth. While the founders of political economy had been radical reformers in their day, championing women's rights, the abolition of slavery, and middle-class interests versus those of the aristocracy, their theories pitted their disciples against labor. As Virginia Woolf's father, Leslie Stephen, remarked: "The doctrine . . . was used to crush all manner of socialist schemes. . . . Political economists were supposed to accept a fatalistic theory, announcing the utter impossibility of all schemes for social regeneration."[43]

For example, when Henry Fawcett, the reform-minded professor of political economy at Cambridge, addressed striking workers, he told them that they were cutting their own throats. Such advice outraged Ruskin, who said, after a builders' strike in 1869, "The political economists are helpless—practically mute; no demonstrable solution of the difficulty can be given by them, such as may convince or calm the opposing parties."[44] Mill was an even more dramatic example than Fawcett. Now a Radical member of parliament, Mill called himself a Socialist, and had championed the Second Reform Act and the right of workers to unionize and strike. Yet Mill's view of the future of the working classes was scarcely less dour than that of Ricardo or Marx. J. E. Cairnes, a professor at University College London who published a famous indictment of slavery as an economic system, echoed Mill's position a few years later:

> The margin for the possible improvement of their lot is confined within narrow barriers which cannot be passed and the problem of their elevation is hopeless. As a body, they will not rise at all. A few, more energetic or more fortunate than the rest, will from time to time escape . . . but the great majority will remain substantially where they are. The remuneration of labor, as such, skilled or unskilled, can never rise much above its present level.[45]

At the heart of Mill's pessimism lay the so-called wages fund theory. According to this theory, ultimately disowned by Mill but never replaced by him, only a finite amount of resources was available to pay wages. Once the fund was exhausted, there was no way to increase the aggregate amount of pay. In effect, the demand for labor was fixed, so that only the supply of labor had any effect on wages. Thus, one group of workers could obtain higher wages only at the expense of lower wages for others.

If unions succeeded in winning a wage rate in excess of the rate of the wages fund, unemployment would result. If the government intervened by taxing the affluent to subsidize wages, the working population would increase, causing more unemployment and even higher taxation. Moreover, the use of taxes to subsidize pay would reduce efficiency by removing competition and the fear of unemployment. Eventually, Mill warned, "taxation for the support of the poor would engross the whole income of the country."[46] Unless the working classes acquired prudential habits of thrift and birth control, the author of a popular American textbook claimed, "they will people down to their old scale of living."[47] In her political economy primer, Millicent Fawcett cited the Corn Law repeal as proof that wages were tethered to a physiological minimum. Referring to the worker, she wrote:

> Cheap food enabled him, not to live in greater comfort, but to support an increased number of children. These facts lead to the conclusion that no material improvement in the condition of the working classes can be permanent, unless it is accompanied by circumstances that will prevent a counter-balancing increase of population.[48]

By the time the Second Reform Act passed however, the theory that wages could not rise in the long run no longer looked tenable, and not only because of the dramatic increase in average pay. The conquest of nature by the railway, steamship, and power loom suggested that society was not yet close to natural limits to growth. The fact that emigrants were prospering abroad and that a middle class of skilled artisans and white-collar workers was shooting up at home contradicted the notion that a mass escape from poverty was ruled out by the biological laws. Poverty that had once appeared to be a natural and near-universal feature of the social landscape began to look more and more like a blemish.

Was there an ingenious mechanism that could lift wages until the average wage sufficed for a middle-class life? Mill acknowledged that the wages fund theory was flawed, but neither he nor his critics could propose a satisfactory alternative. An extraordinary number of Victorian intellectuals—from Charles Dickens, Henry Mayhew, and Karl Marx to John Ruskin and Henry Sidgwick—attempted to fashion one. Since none had so far succeeded, no one could say whether hopes for social betterment really could be reconciled with economic reality, or whether the palpable gains of the 1850s and 1860s were doomed to be reversed. Tories such as Ruskin and Carlyle, an anti-Abolitionist, predicted disaster if the old feudal bonds were not restored. Socialists argued that without sweeping societal changes, the condition of workers was "un-improvable and their wrongs irremediable."[49] The standard-of-living debate, as it became known, boiled down to one question: How much improvement was possible under existing social arrangements?

As he stood before "70 to 80 ladies" in a borrowed Cambridge college lecture hall on a spring evening in 1873, Alfred Marshall's handsome face was lit with an inner flame, and he spoke with great force and fluency without notes. He addressed the women in plain, direct, homely terms as if he were speaking to his sister, urging them to stop "tatting their tatting and twirling their thumbs" and counseled them to resist the demands of their families. Instead he wanted them to get jobs as social workers and teachers like "Miss Octavia Hill." Most of all, he insisted that they learn "what difficulties there are to be overcome, and . . . how to overcome them."[50]

Like his mentor Henry Sidgwick and other university radicals of the 1860s and 1870s, Marshall came to see education as a weapon in the struggle against social injustice, and like other admirers of Mill's *The Subjection of Women*, published in 1869, he considered the educated woman society's principal change agent. For Marshall, the existential problem for women and for the working classes was essentially the same: both lacked the opportunity to lead independent and fulfilling lives. Workers were condemned by low wages to lives of drudgery that prevented all but the most exceptional from fully developing their moral and creative faculties. Middle-class women were condemned by custom to ignorance and drudgery of a different sort. Inspired by the novels of contemporaries such as George Eliot and Charlotte Brontë, Marshall was particularly sensitive to the plight of women who were prevented from developing their intellects and regretted society's loss of their talents. He was convinced that the task of liberating the working classes required the energies of middle-class women as well as a more scientific economics. On the topic of "the intimate connection between the free play of the full and strong pulse of women's thought and the amelioration of the working classes," Marshall was "a great preacher." In an age that celebrated "the angel of the hearth," Marshall taught extension courses for women, acted as an unpaid examiner, and personally financed an essay-writing prize in economics for female students, as well as, later on, contributing a substantial £60 to the construction fund for Newnham Hall, the nucleus of one of Cambridge's first women's colleges. In 1873, Marshall joined Sidgwick, other members of the Grote Club, and Millicent Fawcett—whose sister Elizabeth Garrett was attempting to study medicine—to found the General Committee of Management of the Lectures for Women.[51]

Marshall's lectures focused on the central paradox of modern society: poverty amid plenty. He taught by posing a series of questions: Why hadn't the Industrial Revolution freed the working class "from misery and vice?" How much improvement is possible under current social arrangements based on private property and competition? His answers reflected how far he had distanced himself from the specific assumptions and conclusions of his predecessors. He told the women that philanthropy and political economy were not, as Malthus had supposed and latter-day Malthusians continued to believe, irreconcilable.

Even as he contradicted the conclusions of the founders of political economy, Marshall insisted that the science itself was indispensible. The problem of poverty was far more complicated than most reformers admitted. Economic science, like the physical sciences, was nothing more or less than a tool for breaking down complex problems into simpler parts that could be analyzed one at a time. Intervention based on faulty theories of causes could easily make the problem worse. Marshall cited Adam Smith, David Ricardo, Thomas Malthus, and John Stuart Mill to demonstrate the power of the "engine of analysis" they had constructed, as well as to show

how it had to be improved. Without such a tool, he told them, discovering truths would always be a matter of accident and the accumulation of knowledge with time wholly impossible.

Marshall agreed with Mill that the industrial revolution hadn't liberated him from the tyranny of economic necessity or supplied the material requisites for a "higher life." "Our rapid progress in science and arts of production might have been expected to have prevented to a great extent the sacrifice of the interests of the laborer to the interests of production . . . It has not done so."[52] What he strenuously disputed was the assertion by political economists that it *could* not do so, that the remuneration of labor as such, skilled or unskilled, could never rise much above its present level.[53]

He did not doubt that the chief cause of poverty was low wages, but what caused wages to be low? Radicals claimed that it was the rapacity of employers, while Malthusians argued that it was the moral failings of the poor. Marshall proposed a different answer: low productivity. He cited as evidence the fact that, contrary to Marx's claim that competition would cause the wages of skilled and unskilled workers to converge near subsistence level, skilled workers were earning "two, three, four times" as much as unskilled laborers. The fact that employers were willing to pay more for specialized training or skill implied that wages depended on workers' contribution to *current* output. Or, put another way, that the demand for labor, not only the supply, helped to determine pay. If that was the case, the average wage wouldn't be stationary. As technology, education, and improvements in organization increased productivity over time, the income of the working classes would rise in tandem. The fruits of better organization, knowledge, and technology would, over time, eliminate the chief cause of poverty. Activity and initiative, not resignation, were called for.

Arnold Toynbee the historian later described the significance of Marshall's insight: "Here is the *first great hope* which the latest analysis of the wages question opens out to the laborer. It shows him that *there is another mode of raising his wages besides limiting his numbers*."[54] Workers themselves could influence their own and their children's ability to earn better wages. "The chief remedy, then, for low wages is better education," Marshall told his audience.

He took great pains to demolish Socialists' claim that but for oppression by the rich, the poor could live in "absolute luxury." England's annual income totaled about £900 million, he told the women. The wages paid to manual workers amounted to a total of £400 million. Most of the remaining £500 million, Marshall pointed out, represented the wages of workers who did not belong to the so-called working classes: semiskilled and skilled workers, government officials and military, professionals, and managers. In fact, an absolutely equal division of Britain's annual income would provide less than £37 per capita. Reducing poverty required expanding output and increasing efficiency; in other words, economic growth.

The chief error of the older economists, in Marshall's view, was to not see that man was a creature of circumstances and that as circumstances changed, man was liable to change as well. The chief error of their critics—but, ironically, one that the founders of political economy shared—was a failure to understand the cumulative power of incremental change and the compounding effects of time.

> There are I believe in the world few things with greater capability of poetry in it than the multiplication table . . . If you can get mental and moral capital to grow at some rate per annum there is no limit to the advance that may be made; if you can give it the vital force which will make the multiplication table applicable to it, it becomes a little seed that will grow up to a tree of boundless size.[55]

Ideas mattered when the past was not simply being reproduced but something new was being created. "An organon" or instrument for discovering truths—truths that depended, like all scientific truths, on circumstances—would be an independent force. "The world is moving on," Marshall said, "but the pace at which it moves, depends upon how much we think for ourselves."[56]

• • •

A year later, Marshall was deep in conversation with Henry Sidgwick in
Anne Clough's sitting room on Regent Street, discussing "high subjects"
when he felt someone staring at him.[57] The young woman who sat with
her sewing untouched in her lap looking toward them had a "brilliant
complexion," "deep set large eyes," and masses of mahogany hair "which
goes back in a great wave and is very loosely pinned up behind."[58] Later,
someone said of the twenty-year-old Mary Paley, "She *is* Princess Ida."
The eponymous heroine of the Gilbert and Sullivan opera had "forsworn
the world, / And, with a band of women, shut herself / Within a lonely
country house, and there / Devotes herself to stern philosophies!" Mary
had just broken off her engagement to a handsome but stupid army officer to join a handful of female pioneers seeking a Cambridge education.
Her part in this "outrageous proceeding" was not a rejection of men, or of
the usual terms of marriage. "He who desires to gain their favor must /
Be qualified to strike their teeming brains, / And not their hearts! / They're
safety matches, sir. / And they light only on the knowledge box."[59]

Mary went to one of Marshall's lectures at the coach house at
Grovedodge and listened, enchanted, as he rhapsodized over Kant,
Bentham, and Mill. "I then thought I had never seen such an attractive
face," she confessed, captivated by his "brilliant eyes." She went to a dance
at Marshall's college, and, emboldened by his "melancholy" look, she
asked him to dance "the Lancers." Ignoring his protestations that he didn't
know how, she led him through the complex steps only to be "shocked at
my own boldness."[60] Before long she was among the regular guests at his
"Sunday evening parties" in his rooms at St. John's, where he served her
tea, crumpets, sandwiches, and oranges and showed her his "large collection of portraits arranged in groups of Philosophers, Poets, Artists . . ."

Possibly Mary reminded Marshall of Maggie Tulliver, the intelligent
but math-phobic heroine of George Eliot's *The Mill on the Floss*, who
wanted to learn "the Euclid" like her brother, Tom.[61] At the time, Eliot's
novel was Marshall's favorite. Meeting Mary Paley and her best friend,
Mary Kennedy, in the street one day, Marshall proposed—not marriage,
but something more outrageous. The young professor wanted his two best

students to take the Moral Sciences Tripos, the final examination in political economy, politics, and philosophy that male undergraduates had to take to get a degree. This was a far more ambitious project than acquiring "general cultivation" by attending lectures in literature, history, and logic, Mary's original object in coming to Cambridge.

The suggestion was also bolder than anything proposed by other education reformers whose main interest lay in raising the level of secondary-school teaching. "Remember, so far you have been competing with cart horses," Marshall warned, "but for the Tripos it will be with racehorses." He promised that he and Sidgwick would coach her. According to Mary Kennedy, "He explained that this would mean at least three years' study, specializing in one or two subjects. We accepted the challenge lightly, not realizing what we were undertaking."

Like Marshall, the young woman who would accept the challenge came from a strict evangelical household. Mary Paley's great-grandfather was William Paley, the archdeacon of Carlisle and author of *The Principles of Moral and Political Philosophy*. Mary's father was the rector of Ufford, near Stamford, about forty miles northwest of Cambridge. A "staunch Radical" who opposed fox hunting, horse racing, and High Church ritual, he refused to talk to neighboring clergymen and forbade his daughters Dickens and dolls. Mary recalled, "My sister and I were allowed dolls until one tragic day when our father burnt them as he said we were making them into idols and we never had any more."

Mary's father was nonetheless a more tolerant, better-educated, and more affluent man than William Marshall. Mary grew up in a "rambling old house, its front covered with red and white roses and looking out on a lawn with forest trees as a background, and a garden with long herbaceous borders and green terraces." The Paley household was a hive of activity: rounders, archery, croquet, excursions to London, summer holidays in Hunstanton and Scarborough. "We had a father who took part in work and play and who was interested in electricity and photography," Mary recalled. Her mother "was full of initiative and always bright and amusing." In 1862 Mary was taken to London to tour the Second Great Exhibition. Although Charles Dickens was taboo, Mary read *Arabian Nights, Gulliver's*

Travels, the *Iliad* and the *Odyssey*, Greek and Shakespearean plays, and the
novels of Sir Walter Scott, also favorites of Marshall's.

When the Cambridge Higher Local Examination for Women over
Eighteen was established in 1869, Tom Paley encouraged Mary to take
it over the objections of her mother. After she succeeded brilliantly and
broke off her engagement to the army officer, her father allowed her to
go to Cambridge to live "when such a thing had never been done before."
Anne Jemima Clough, a friend of Sidgwick's and one of the leaders of the
women's education movement, was opening a residence for a handful of
female students. Mary later wrote, "My father was proud and pleased and
his admiration for Miss Clough overcame his objections to sending his
daughter to Cambridge (in those days an outrageous proceeding)." [62]

In October 1871, Mary joined Miss Clough and four other young
women at 74 Regent Street. The Cambridge community was wholly un-
prepared for coeducation. Since mixed classes were "improper," sympathetic dons had to be recruited to repeat their regular lectures separately
for the women, and Miss Clough, as chaperone, had to sit through them
all. The "strong impulse towards liberty among the young women attracted by the movement" and the "unfortunate appearance" of the pretty
ones were chronic sources of anxiety. Mary, who was just entering her
"pre-Raphaelite period" and had papered her rooms in William Morris
designs, was especially troublesome. She dressed as if she were a figure in
an Edward Burne-Jones painting, in sandals, capes, and flowing gowns. An
amateur watercolorist, she favored jewel tones and once covered her tennis
dress with Virginia creeper and pomegranates.

Mary began to go regularly. Earnest as well as artistic, with a quick
facility for "curves," the graphs that Marshall employed to illustrate the
interactions of supply and demand, Mary surprised herself by winning
the essay prize. She was thrilled by Marshall's bold proposal that she take
the Tripos, and the long comments he wrote on her weekly papers in red
ink became "a great event."

Mary Paley took the Moral Sciences Tripos in December 1874. Until
the eve of the examination, it was unclear whether the university examin-

ers would be willing to let her sit for it. One was considered "very obdurate." Although they grudgingly agreed to grade her examination, they refused to grant her the highest mark. "At the Examiners' Meeting there was at that time no chairman to give a casting vote, and as two voted me first class and two second class I was left hanging, as Mr. Sidgwick said, 'between heaven and hell,'" she later recalled. Still, her triumph turned Paley into a local celebrity.

Her time at Cambridge seemingly having run out, Mary returned to the family home in Ufford. There she promptly organized a series of extension lectures for women—"off my own bat!"—in nearby Stamford. She also agreed, at the suggestion of a Professor Stuart at Cambridge, to write a textbook on political economy for use in the extension courses. Then she got a letter from Sidgwick asking whether she could take over Marshall's economics lectures at Newnham, where Miss Clough had assembled about twenty students.

At thirty-two, Marshall was one of the "advanced liberals" at Cambridge University. He wore his hair fashionably long, sported a handlebar mustache, and no longer dressed like a buttoned-up young minister. He had joined the recently founded Cambridge Reform Club and read the *Bee Hive,* a radical labor magazine.

In the spring of 1874, a farmworkers' strike provoked a bitter quarrel between radicals and conservatives at Cambridge. Trade unions were then relatively novel, having only just been legalized. The National Agricultural Laborers' Union, a radical new organization under the leadership of Joseph Arch, had sprung up in dozens of East Anglian villages the previous fall. The laborers demanded higher wages and shorter hours as well as the franchise and reform of the land laws.[63] Strikes erupted all around Cambridge. Determined to "crush the rebellion," farmers banded together in "Defense Committees," firing and evicting men with union cards and importing scab labor from as far away as Ireland. The Tory *Cambridge Chronicle* suggested that the farmers "do not make a stand so much against an increase of wage as against the cunning tactics and insufferable dicta-

tion of the union through demagogue delegates."[64] By mid-May, the lockout was two and a half months old and had become the subject of national
controversy.

At the university, where a large subscription had just been undertaken
for famine victims in Bengal, opinion was sharply divided. Middle-class
sympathies for the plight of the laborers had been awakened by a number
of inquiries, most notably a Royal Commission report by the bishop of
Manchester, who had exposed the long hours, low wages, horrific accidents, and diets of "tea kettle broth, dried bread and a little cheese" endured by agricultural workers.[65] During the lockout, the *Times* of London
ran stories calculated to horrify Victorian readers, including one description of a cottage whose single bedroom was shared by "the laborer, and his
wife, a daughter aged 24, and a son aged 21, another son of 19, and a boy
of 14, and a girl of 7."[66] Novelists seized on the subject as well. In George
Eliot's *Middlemarch,* which had appeared three years earlier, Dorothea
Brooke tells her uncle, a well-to-do landlord, that she cannot bear the
"simpering pictures in the drawing-room . . . Think of Kit Downes, uncle,
who lives with his wife and seven children in a house with one sitting-
room and one bedroom hardly larger than this table!—and those poor
Dagleys, in their tumble-down farmhouse, where they live in the back-
kitchen and leave the other rooms to the rats! That is one reason why I did
not like the pictures here, dear uncle."[67]

Among conservatives, however, the unrest raised the specter of
the Bread Riots of 1816–17 and the burning of hayricks in the 1830s.
Most opposed the idea of unionization on principle. In the spring
a leading member of the university community, who was of "recognized social position . . . occupying an influential position in one of
[Cambridge's] . . . colleges," wrote several lengthy "Notes of Alarm" in the
Cambridge Chronicle urging the farmers to stand fast. He labeled the union
leaders "professional mob orators" and their liberal sympathizers "sentimental busybodies." The writer—possibly a Cambridge don named William Whewell—signed himself only "CSM," an acronym probably chosen
to provoke his liberal opponents because it stood for Common Sense Morality. On the matter of wages and unionization, CSM invoked the laws of

political economy, claiming, "It is simply a question of supply and demand, and ought to have been allowed to settle itself on ordinary principles without the interference of paid agitators and demagogues."[68]

The overflowing crowd of union supporters that squeezed into the Barnwell Workingmen's Hall on Cambridge's scruffy north side on Tuesday, May 11, 1874, was thus somewhat bemused to find an unlikely set of allies standing on the stage clad in caps and gowns. One of the leaders, the fiery George Mitchell, confessed, amid much laughter, that "when he saw all those gentlemen with their wide-awake hats and tippets he thought he was going to have some put on him."[69] Sedley Taylor, a former Trinity College fellow and prominent reformer, spoke first, proposing a resolution condemning the farmers' efforts to break the union as "prejudicial to the general interests of the country," delivering a broadside at his fellow collegian CSM in the process.

Then it was Marshall's turn. Seconding a motion put forward by a dissident farmer supporting the locked-out laborers, he called for donations: "Let us sympathize with our hearts and with our purses."

Addressing the farmworkers, Marshall denied that political economy could "direct decisions of moral principle," which it must instead "leave to her sister, the Science of Ethics." Writing in the *Bee Hive,* he argued that "political economy is abused when any one claims for it that it is itself a guide in life. The more we study it the more we find cases in which man's own direct material interest does not lie in the same direction as the general well being. In such cases we must fall back on duty."[70]

The following Saturday, the *Cambridge Chronicle* dismissed Marshall's speech as "ingenious sophistry." In fact, he had successfully demonstrated why labor markets do not always produce fair wages, and why unions can lead to greater efficiency as well as equity. He'd "been asked to speak of the laws of supply and demand," Marshall began. He poured scorn on the union's opponents who held wages were at their "natural level" because, if they weren't, other employers would have offered the workers more, and if a worker's "wages be raised artificially they will come down again." This was Ricardo's iron law of wages, accepted even by many who sympathized with the plight of the workers. The argument was "excellent," Marshall admitted, but the assumptions false. No farmer would offer a neighbor's hired hands more to come and work for him. What's more, higher wages would make the workers more productive by allowing them to be better fed. Admitting that "unions have their faults," Marshall said that "a union gives men interests and sympathies beyond the boundaries of their parish; it will cause them to feel their need of knowledge, and to vow that their sons shall be educated . . . Wages will rise . . . poor rates will dwindle . . . England will prosper."[71]

Despite the support of the university and much of the media, the strike ultimately failed. The farmers held out by acquiring more machinery and employing more boys and girls. When the strike fund ran out in early June, the union called on the workers to return to the fields. Marshall took from the episode that new ideas would prevail over old doctrines only after a carefully plotted, patient campaign to win the hearts and minds of practical men.

Five weeks out of New York City and bound for San Francisco, Marshall stared down on the Horseshoe Falls with a frown. From the Goat Island suspension bridge where he stood, the cataract looked nowhere near as mighty as his Baedeker guide had promised. As a mathematician, he knew that perspective was to blame and engaged in some mental calculations to reassure himself that the falls were truly as colossal as advertised. But the numerical exercise did little to dispel his feeling of having been badly let down. "Niagara is a great humbug," he wrote to his mother on July 10, 1875. "It takes longer for a man to discover how much greater Niagara is than it seems than it does to discover that an Alpine Valley which appears to be only a mile broad is really six miles broad."[72]

Marshall had come to America to study its social and economic landscape. He had left Manhattan on a paddle steamer headed for Albany. In a letter, he recalled how "disgusted and savage" Alexis de Tocqueville had been forty years earlier when he discovered that the finest of the "villas built in Greek style of marble, shining from the banks of the Hudson" were actually made of wood. He, by contrast, "did not find anything like as much sham as I expected."[73]

Indeed, everywhere Marshall looked, he seemed to discover more, not less, than met the eye: American architects displayed "daring & strength," their buildings being of "uniform thoroughness & solidity."[74] An "American drink called 'mint-julep'" was "luxurious." American preachers gave sermons that were "way out of sight ahead of us," having achieved "startling improvements" on Anglican liturgy.[75] American workers were full of "go."[76] As he reported to the Moral Sciences Club on his return to Cambridge in the fall, "I met no man or woman in America whose appearance indicated an utterly dull or insipid life."[77] By the time Marshall reached Cleveland in mid-July, he was convinced that "nine Englishmen out of ten would be themselves more happy & contented in Canada than in the U.S.; though I myself if I had to emigrate should go to the U.S."[78]

Marshall's magnum opus, *Principles of Economics,* would not appear for another fifteen years, but he had already worked out the chief tenets of his "new economics"—an alternative to both the old laissez-faire doctrines of Smith, Ricardo, and Mill and the newly ascendant Socialist gospels of Marx. He had spent a decade "laying the foundations of his subject but publishing nothing."[79] His travels in America gave him confidence that he was on the right track.

Marshall's relations had scoffed at his plan to use a £250 legacy from the same uncle who had financed his university education to tour the United States. He justified himself by saying that he was gathering material for a treatise on foreign trade. While this was perfectly true, the economic theorist John Whitaker observes that his actual purpose was broader, part of a growing, "almost obsessive attempt to apprehend in all its aspects an ever-changing economic reality."[80] Like other European observers, including Tocqueville, Marshall thought of the United States as a great social laboratory. Dickens, William Makepeace Thackeray, and Trollope had been occupied by old questions, now settled, of democracy, slavery, and the survival of the union. Marshall wanted to know where the rise of industry, the growth of global commerce, and the decline of traditional morality were leading. These were advancing more rapidly in America than anywhere else. "I wanted to see the history of the future in America," he told an audience when he returned to Cambridge.[81]

Marshall sailed to America during the biggest transatlantic tourism boom in history. Sales of the most popular North American guide were climbing toward the half-million mark. The North Atlantic was now a virtual highway of the sea. No fewer than ten steamship companies offered weekly departures from Liverpool to New York, and English travelers were advised to book berths as much as a year in advance.[82] Marshall's trip aboard the SS *Spain,* one of the fastest and most luxurious of the big liners, took a mere ten days, in contrast to the miserable three-week crossing Dickens had endured in 1842. Travel in America was expensive, owing to the immense distances. Marshall had to budget £60 a month versus £15 a month when he spent summers climbing in the Alps. But afterward, according to Mary, he felt that "he had never spent money so well. It was not so much what he learnt there as that he got to know what things he wanted to learn."[83]

His experiences convinced him that "economic influences play a larger part in determining the higher life of men and women than was once considered." In particular, he believed, "there are no thoughts or actions, or feelings, which occupy a man and which thus have the opportunity of forming the man . . . as those thoughts and actions and feelings which make up his daily occupation."[84] He spent some of his time in churches and drawing rooms, especially in Boston, where he met leading American intellectuals, including the poet Ralph Waldo Emerson and the art historian Charles Eliot Norton. He lingered for several days at communes run by Shakers and disciples of Robert Owen in New England. But mostly he toured factories, filling notebooks with interviews with businessmen and workers and drawings of machinery. At Chickering and Sons piano factory near Boston, he observed that "care & judgment were required from many of the workers in a very high degree" and that the workers there had "able, almost powerful & artistic faces." On a visit to an organ factory, he wondered whether "the work of each individual being confined to a very small portion of the whole operation" did not "prevent the growth of intelligence?"[85] He found that it did not.

The business traveler of that time was always something of a tourist. Marshall was no exception. He could not resist the lure of the recently completed transcontinental railroad. In his hotel in Niagara, he plotted his westward route on an advertising map provided by the Union Pacific, marking it with pinpricks so that his mother back home in London could follow his progress toward San Francisco by holding the map up to a light.

Chicago was the best place to catch a train for the Pacific coast. The new railway system was like a giant hand whose palm lay atop the Great Lakes and whose fingers stretched all the way to Seattle, Portland, San Francisco, and, in the case of the two southernmost routes, Los Angeles. Most travelers took the North Western from Chicago due west across Illinois and Iowa to Council Bluffs. Marshall took the Great Northern line to St. Paul and then sailed back down on a Mississippi riverboat, the kind "more famous for their propensity to blow up than for the magnificence of their fittings."[86] He met up with the North Western at the Iowa border and was in Council Bluffs a day later. From there he crossed the river to Omaha and transferred to the Union Pacific train. From Omaha it was a straight shot west to Cheyenne and Granger, in Wyoming, where the line dipped down toward Ogden, Utah; Reno; and Sacramento before making the final 125-mile jog south to San Francisco. In Cheyenne, Marshall boarded a stagecoach for a twenty-four-hour side trip to Denver. In Ogden, he stopped to explore the Mormon capital, Salt Lake City. On the return trip, he got off in Reno for a look at "the wild population of Virginia City." He was conscious throughout of witnessing something extraordinary and unprecedented. From his railway car he was seeing what another young Briton had earlier described as "the unrolling of a new map, a revelation of a new empire, the creation of a new civilization."[87]

Marshall was bowled over by the constant motion he witnessed. "Many things have changed since [Tocqueville's] time . . . many things which were nearly stationary then are not stationary now," he wrote in a letter home.[88] The first thing to catch his eye after he checked in at the Fifth Avenue Hotel was "a steam lift which *without ever stopping* from 7 a.m. until midnight goes up & down [emphasis his]." He was captivated by the lobby's unmanned telegraph machine spewing paper ribbons of stock quotations. Business travelers staying uptown "are as well posted as if they were on the Exchange itself," he wrote.[89]

Mobility was the preeminent fact of American life, Marshall decided. It wasn't just the railway and telegraph, the successive waves of new immigrants, or the movement of the population from the manufacturing centers of the Northeast to the "mushroom towns" of the West, sprouting so fast that one "can only suppose that, the soil being so fruitful, buildings grow spontaneously."[90] The most interesting freedom of motion was economic, social, and psychological. Marshall was astonished by ordinary Americans' readiness to leave family and friends for new towns, to switch occupations and businesses, to adopt new beliefs and ways of doing things. He reported, "If a man starts in the boot trade and does not make money so fast as he thinks he ought to do, he tries, perhaps, grocery for a few years and then he tries books or watches or dry goods." He was delighted by the independence of young people: "American lads . . . abhor apprenticeships . . . The mere fact of his being bound down to a particular occupation is sufficient in general to create in the mind of an American youth that he will do something else as soon as he has the power."[91]

Americans' welcoming attitude toward growing urbanization also struck him powerfully: "The Englishman Mill bursts into unwonted enthusiasm when speaking . . . of the pleasures of wandering alone in beautiful scenery," he noted dryly, adding that "many American writers give fervid descriptions of the growing richness of human life as the backwoodsman finds neighbors settling around him, as the backwoods settlement develops into a village, the village into a town, and the town into a vast city."[92]

Like his favorite novelists, Marshall was less interested in the material and technological advances, impressive as these were, than in their consequences for how people thought and behaved. What guarantee was there that individual choices added up to social good? Would all the up and down movement of individuals and the attendant loosening of traditional ties lead, as pessimists such as Marx and Carlyle predicted, to social chaos? Or did mobility imply a "movement towards that state of things to which modern Utopians generally look forward." That was the question.[93]

Marshall's visceral reactions put him squarely on the other, optimistic side. In Norwich, Connecticut, he went on an evening drive with a Miss

Nunn, who told him she was prepared to take the reins and wound up steering. Marshall found the experience "very delicious." He observed that young American women are "mistresses of themselves . . . [with] thorough freedom in the management of their own concerns." Such freedom, he admitted, "would be regarded as dangerous license by the average Englishman," but he found it "right and wholesome."[94]

The absence of rigid class distinctions delighted him. When a clerk in a hat shop removed the bowler Marshall was wearing and tried it on his own head in order to gauge the correct size, Marshall noted approvingly, "My friend was such a perfect democrat that it did not occur to him that there was any reason why he should not wear my hat: his manner was absolutely free from insolence. May the habit become general!"[95] When he reached California, he was pleased to report that the farther west he traveled, the more American society resembled its egalitarian ideal. "I returned on the whole more sanguine with regard to the future of the world than when I set out," he noted.

Striking a prophetic note, he envisioned a new type of society:

In America, mobility was creating an equality of condition . . . Where nearly all receive the same school education, where the incomparably more important education which is derived from the business of life, however various in form it be, yet is for every one nearly equally thorough, nearly equally effective in developing the faculties of men, there cannot but be true democracy. There will of course be great inequalities of wealth; at least there will be some very wealthy men. But there will be no clearly marked gradation of classes. There will be nothing like what Mill calls so strongly marked line of demarcation between the different grades of laborers as to be almost equivalent to the hereditary distinction of caste.

Explaining how individual choices might add up to social good—the very thing that Carlyle denied was possible—Marshall defined two types of moral education. One was characteristic of England, where, he claimed, "the peaceful molding of character into harmony with the conditions by which it is surrounded, so that a man . . . will without conscious moral effort be impelled on that course which is in union with the actions, the sympathies and the interests of the society amid which he spends his life." In America, by contrast, mobility had opened up a second route to moral evolution, namely, "the education of a firm will by the overcoming of difficulties, a will which submits every particular action to the judgment of reason."[96]

Most Victorian social commentators, including Karl Marx, feared that the industrial system was not merely destroying traditional social relations and livelihoods but deforming human nature through "ignorance, brutalization, and moral degradation."[97] In America, Marshall saw another possibility: "It appears to me that on the average an American has the habit of using his own individual judgment more consciously and deliberately, more freely and intrepidly, with regards to questions of Ethics than an Englishman uses his."

Marshall seemed to be talking about mankind in general, but he was also talking about himself. *He* had developed a firm will by overcoming all sorts of difficulties—a tyrant of a father, genteel poverty, and the oppressive strictures of class. *He* had broken with authority—by losing his religious belief and defying his father's wishes that he enter the ministry. Now he felt that his own independence would lead not to his downfall but to great things. What he witnessed in America filled him with hope. "Such a society may degenerate into licentiousness and thence into depravity. But in its higher forms it will develop a mighty system of law, and it will obey law . . . Such a society will be an empire of energy."[98]

"I have been rather spoilt" when it comes to "go" and a "strong character" in women," Marshall had written in a letter from America. In another, he described his "riveting evening" with Miss Nunn, confessing that he found her naïveté "mingled with enterprise" charming. But he added that "for steady support I would have the strength that has been formed by daring and success."[99] Apparently he was thinking of Mary Paley, who had triumphed over the Tripos in his absence.

When they got engaged on his return to Cambridge, Marshall was thirty-four and Mary twenty-six. He was a rising star of the "New Economics." She was a college lecturer. Marshall's view of marriage was inspired by intellectual partnerships such as those of George Eliot and George Lewes and Thomas and Jane Carlyle. "The ideal of married life is often said to be that husband and wife should live for each other. If this means that they should live for each other's gratification it seems to me intensely immoral," Marshall wrote in an essay. "Man and wife should live, not for each other but with each other for some end." [100] For Mary, who had entered her first engagement "out of boredom," this was a thrilling vision. Like the other unusual, idiosyncratic Victorian marriages Phyllis Rose describes in *Parallel Lives: Five Victorian Marriages,* the secret of Alfred Marshall and Mary Paley's alliance lay in their "telling the same story." [101] The couple immediately decided to make Mary's textbook a joint project and spent most of their engagement working on it.

They were married at the Parish Church in Ufford, next to the "rambling old house, its front covered with red and white roses," where Mary had grown up. Mary wore no veil, only jasmine in her hair. In a gesture that proclaimed their untraditional views and high expectations, bride and groom contracted themselves out of the "obey clause." [102]

By marrying, Marshall forfeited his fellowship at St. John's. He and Mary flirted briefly with the notion of teaching at a boarding school, but when the principalship of a newly founded redbrick college in Bristol— the first experiment in coeducation in Britain—suddenly became vacant, they leapt at the opportunity. When they moved to Bristol in 1877, Mary had a tennis court installed and most of the rooms papered with Morris while Marshall chose the secondhand furniture and piano. But she was soon back in the classroom, lecturing on economics and tutoring women students.

Underwritten by Bristol's business community, University College was to provide "middle and working class men and women with a liberal education." [103] Though strapped for funds, the college managed, during the Marshalls' tenure, to offer day and evening classes to some five hundred students, sponsor public lectures in working-class neighborhoods, provide technical instruction to textile workers, and run a work-study program

jointly with local businesses for engineering students. Marshall's administrative duties were heavy and so was his teaching load. His regular classes,
attended by a mix of small businessmen, trade unionists, and women,
were "less academic than those at Cambridge . . . a mixture of hard reasoning and practical problems illuminated by interesting sidelights on all
sorts of subjects," a student recalled.[104] Marshall "spoke without notes and
his face caught the light from the window while all else was in shadow. The
lecture seemed to me the most wonderful I had ever heard. He told of his
faith that economic science had a great future in furthering the progress
of social improvement and his enthusiasm was infectious."[105] The couple
continued to work on *The Economics of Industry* most afternoons, took
long walks, and played many games of lawn tennis. One friend referred to
"their perfect happiness."[106]

Marshall later said that reading Marx convinced him that "economists
should investigate history; the history of the past and the more accessible
history of the present."[107] But it was Dickens and Mayhew who inspired
him to go into factories and industrial towns to interview businessmen,
managers, trade union leaders, and workers. "I am greedy for facts," he
used to say.[108] He wanted to write for men and women engaged in the "ordinary business of life."[109]

He was convinced that he would have to blend theory, history, and
statistics, as Marx had done in *Das Kapital*. But he was instinctively aware
that his audience would require useful practical conclusions and a gener-
ous sprinkling of direct observation. He was too much of a scientist to
theorize without verifying facts, or to rely on secondhand descriptions.

Marshall made a commitment to study the particulars of every
major industry. He gathered data on wage rates by occupation and skill
level. He paid a great deal of attention to Mill's "arts of production"[110]—
manufacturing techniques, product design, management—although he
admitted that the constant effort of business owners to improve their
products, production methods, and suppliers was hard to capture in
formal theories. He was particularly interested in how the family-owned,
privately held firm functioned versus the increasingly important joint

stock company or corporation. Marshall participated in commissions and learned societies and sat on the board of a London charity, carried on a huge scientific correspondence, and, with Mary as an active partner, devoted several weeks each summer to fieldwork.

On one such quest, Mary's notes refer to "14 different towns, mines, iron and steel works, textile plants, and [the] Salvation Army."[111] The itinerary was extraordinarily ambitious: Coniston copper mines, Kirby slate quarries, Barrow docks, iron and steel works, Millom iron mines, Whitehaven coal mines close to the sea, Lancaster, and Sheffield. Marshall invented a device for organizing and retrieving information from his personal database. His "Red Book" was a homemade notebook sewn together with thread. Each page contained data on a variety of topics, ranging from music to technology to wage rates, arranged in chronological order. Marshall had only to stick a pin through one of the points on a page to see what other developments had occurred simultaneously.

In contrast to the majority of Victorian intellectuals, Marshall admired the entrepreneur and the worker. Carlyle, Marx, and Mill considered modern production to be an unpleasant necessity, labor to be degrading and debilitating, businessmen to be predatory and philistine, and urban life to be vile. Mill considered Communism superior to competition in every respect but two (motivation and tolerance for eccentricity) and looked forward to a stationary, Socialistic state in the not very distant future. But none of these intellectuals could claim the familiarity with business and industry that Marshall was acquiring. Of course, as Burke's phrase "drudging through life" implied, much of human labor had and was having such effects. But, once again, Marshall's reliance on firsthand observation suggested that at least some work in modern firms expanded horizons, taught new skills, promoted mobility, and encouraged foresight and ethical behavior, not to mention provided the savings to go to school or into business. What was more, he observed, that sort of work was growing while the other was becoming less common. In short, the business enterprise could be and often was a step toward controlling one's destiny.

Although Dickens is often thought of as a chronicler of the industrial revolution, almost the only factory scene in Dickens is phantasmagorical. The Coketown factory in *Hard Times* is a Frankenstein, seen only from a distance, that turns men into machines and re-creates the natural and social environment in its own monstrous image; noisy, dirty, monotonous, its air and water poisoned.

> It was a town of red brick, or of brick that would have been red if the smoke and ashes had allowed it; but, as matters stood it was a town of unnatural red and black like the painted face of a savage. It was a town of machinery and tall chimneys, out of which interminable serpents of smoke trailed themselves for ever and ever, and never got uncoiled. It had a black canal in it, and a river that ran purple with an ill-smelling dye, and vast piles of building full of windows where there was a rattling and a trembling all day long, and where the piston of the steam-engine worked monotonously up and down, like the head of an elephant in a state of melancholy madness.[112]

Coketown is inhabited by an army of "people equally like one another, who all went in and out at the same hours, with the same sound upon the same pavements, to do the same work." Significantly, Dickens imagines that inside the factory they "do the same work" and that "every day was the same as yesterday and to-morrow, and every year the counterpart of the last and the next." Production, in other words, involves never creating anything new.

Marx's description of the factory in *Das Kapital* stresses the same features as Dickens's but lacks all detail, not surprising given that Marx had never been inside even a single one. Again, men are transformed into a "mere living appendage" of the machine, work becomes "mindless repetition," and automation "deprives the work of all interest."[113]

Marshall's descriptions of factories and factory life are more specific, nuanced, and varied. He spends hours observing. He records manufacturing techniques and pay scales and layouts. He questions everyone, from the owner to the foremen to the men on the shop floor. When he encounters the same problematic phenomenon as Dickens or Marx—the effects of the assembly line on workers—he doesn't necessarily draw the same inferences.

> The characteristic of the firm is the way in which every operation is broken up into a great number of portions, the work of each individual being confined to a very small portion of the whole operation. Does this prevent the growth of intelligence? I think not . . . If a man has no brains we get rid of him: There is plenty of opportunity for this in consequence of the fluctuations of the market. If a man has some brains, he stays on at his work; but if he has any ambition, he must get to know all that goes on in the shop in which he is working: otherwise he has no chance of becoming foreman of that shop . . . Most improvements in detail are made by the foremen of the several shops: & improvements on a very large scale are made by a man who does nothing else . . . Their improvements were in small details as regards manufacture e.g. numerous contrivances for securing that certain parts should be airtight, that certain others should work easily. The Englishman had invented the harp stop.[114]

For Dickens and Marx, firms existed to control or exploit the worker. For Mill they existed solely to enrich their owners. For Marshall, the business firm was not a prison. Management wasn't just about keeping the prisoners in line. Competing for customers (or workers) required more than mindless repetition. Marshall's business enterprises were forced to evolve in order to survive. Of course, Marshall did not deny that businessmen pursued profits. His point was that to make profits competitive, firms had to generate enough revenue to still have something left over after paying workers, managers, suppliers, landlords, taxes, and so on. To do that, managers had to constantly seek out ways to do a little more with the same or fewer resources. In other words, higher productivity, the long-run determinant of wages, was a by-product of competition.

The British publisher Macmillan & Co. brought out *The Economics of Industry* in 1879. A slim volume purporting to contain nothing new and written in simple and direct prose suitable for a primer, it contained the essentials of Marshall's New Economics. Its message was summarized in the following passage:

> The chief fault in English economists at the beginning of the century was not that they ignored history and statistics . . . They regarded man as so to speak a constant quantity and gave themselves little trouble to study his variations. They therefore attributed to the forces of supply and demand a much more mechanical and regular action than they actually have; But their most vital fault was that they did not see how liable to change are the habits and institutions of industry.[115]

Marshall's obsessive effort to understand how businesses worked led to his most important discovery. The economic function of the business firm in a competitive market was not only or even primarily to produce profits for owners. It was to produce higher living standards for consumers and workers. How did it do this? By producing and distributing more goods and services of better quality and at lower cost with fewer resources. Why? Competition forced owners and managers to constantly make small changes to improve their products, manufacturing techniques, distribution, and marketing. The constant search to find efficiency gains, economize on resources, and do more with less resulted over time in doing more with the same or fewer resources. Multiplied over hundreds of thousands of enterprises throughout the economy, the accumulation of incremental improvements over time raised average productivity and wages. In other words, competition forced businesses to raise productivity in order to stay profitable. Competition forced owners to share the fruits of these efforts with managers and employees, in the form of higher pay, and with customers, in the form of higher quality or lower prices.

The implication that business was the engine that drove wages and living standards higher ran counter to the general condemnation of business by intellectuals. Even Adam Smith, who famously described the benefits of competition in terms of an invisible hand that led producers to serve consumers without their intending to do so, had not suggested that the role of butchers, bakers, and giant joint stock companies was to raise living standards. Although Karl Marx had recognized that business enterprises were engines of technological change and productivity gains, he could not imagine that they might also provide the means by which humanity could escape poverty and take control of its material condition.

A serious crisis followed the publication of the Marshalls' book. Marshall was diagnosed with a kidney stone in the spring of 1879. Surgery and drugs were not options at that time. His doctor said, "There must be no more long walks, no more games at tennis, and that complete rest offered the only chance of cure," Mary recalled later. "This advice came as a great shock to one who delighted so in active exercise." [116] The painful, debilitating condition revived Marshall's old fears of impending annihilation, still lurking from childhood. Only a few weeks earlier, he had spent a vacation hiking alone on the Dartmouth moors. Now he had become a housebound invalid who took up knitting to pass the time. A Bristol acquaintance recalled seeing Marshall and thinking that he must be seventy or so:

> He . . . looked to me very old and ill. I was told he had one foot in the grave and I quite believed it. I can see him now, creeping along Apsley Road . . . in a great-coat and soft black hat . . . The next time I saw him was . . . in 1890 . . . I was astonished to find him apparently thirty or forty years younger than I remembered him a dozen years before. [117]

It made him more dependent on Mary and caused him to cast her ever more into the role of nurse rather than intellectual companion. Illness concentrated his mind. Marshall always had a tendency toward writer's block. Now he realized he had to focus his energies and get on with his book. His hopes for writing a work that would eclipse Mill's (and perhaps also Marx's)—a synthesis of new theory and freshly distilled reports from the real world—were matched by fears that he was not up to the task. As his vision grew in scope and complexity, he grew proportionately less satisfied with what he had written. He had decided to drop plans to publish his volume on trade well before his illness flared. "I have come to the conclusion that it will never make a comfortable book in its present shape," he wrote in the summer of 1878.[118] And he quickly grew to dislike the book he had written with Mary. But in 1881, on a rooftop in Palermo, Sicily, he began to compose *Principles of Economics*.

Of all the panaceas advanced during the Great Depression of the early 1880s, the American journalist Henry George's land tax attracted by far the most popular attention and support. George's best seller, *Progress & Poverty*, had made him an instant celebrity, and his lectures drew huge crowds. George's premise was that poverty was growing faster than wealth and that landlords were to blame. He claimed that landlords were collecting fabulous incomes not for rendering a service to the community but merely because they were lucky enough to own real estate. What was more, rising rents were depressing profits and real wages by depriving businessmen of needed investment funds. Having identified rental income as the cause of poverty, he proposed a massive tax on land as a cure. The land tax would not only eliminate the need for all other taxes, he claimed. It would also "raise wages, increase the earnings of capital, extirpate pauperism, abolish poverty, give remunerative employment to whoever wishes it, afford free scope to human powers, lessen crime, elevate morals and taste and intelligence, purify government, and carry civilization to yet nobler heights."[119]

Marshall was still working on *Principles* when he was drawn once again into the long-simmering standard-of-living controversy. The early 1880s, a period of financial and economic crisis, witnessed a resurgence of radicalism and demands for social reform, as well as growing skepticism about the extent to which economic growth was benefiting the majority of citizens. The term unemployment was coined during the recession that followed the Panic of 1893 during a heated debate over whether real wages were rising or falling in the long run.

At issue in the debate was the dominant effect of competition. Did competition result in a race to the bottom in which employers matched one another's wage cuts? Or was it the case, as optimists insisted, that competition put pressure on companies to make constant efforts to increase efficiency and push up the average level of productivity and wages while reducing the number of poor?

The first formal confrontation between Marshall and Henry George took place at the Clarendon Hotel in Oxford in 1884.[120] Catcalls, clapping, and hissing repeatedly drowned out the debaters. At one point, an undergraduate felt it necessary to primly remind the chairman that "ladies were present." By eleven o'clock, the uproar was so deafening that George declared the meeting to be "the most disorderly he had ever addressed" and refused to answer any more questions. Amid "great noise" and groans of "Land Nationalization" and "Land Robbery," the meeting "was brought to a rather abrupt conclusion."

If Marshall's support for the agricultural lockout in 1874 signaled his rejection of the "dogmas" of classical economy, his confrontation with George a decade later showed that he also objected to trendy new dogmas.

On other occasions when he had criticized George's proposal to cure poverty with a tax on land, Marshall had called George a "poet" and praised "the freshness and earnestness of his view of life." But at the Clarendon, Marshall was decidedly less polite, accusing George of using his "singular and almost unexampled power of catching the ear of the people" to "instill poison into their minds." By "poison," he meant George's cure-all for poverty.

In his Bristol lectures, Marshall stuck to his stated intention to "avoid talking very much about George: but to discuss his subject." George's subtitle included an inquiry into "the increase of want with the increase of wealth," Marshall said. "But are we sure that with the increase of wealth want has actually increased? . . . Let us then enquire what the facts are of the case."[121]

Citing statistical evidence—much of it collected in the Red Book that he and Mary had compiled—Marshall argued that only the "lowest stratum" of the working classes were being pushed downward and that that stratum was far smaller—less than half the size, in proportion to the population—than it had been earlier in the century. As for the working classes as a whole, their purchasing power had tripled. "Nearly one half of the whole income of England goes to the working classes . . . [So] a very large part of all the benefit that comes from the progress of invention must fall to their share." [122]

Marshall drew on his growing command of economic history. He was confident that, whatever the vices of the current age, they paled in comparison to the past. "The working classes are in no part of the world, except new countries, nearly as well off as they are in England." What makes Marshall's optimism all the more noteworthy is that he was speaking during what historians would later call the Great Depression.

In his second lecture, Marshall challenged George's contention that employers who paid low wages were to blame for poverty. For one thing, employers could not set the price of labor any more than they could dictate the price of cotton or machinery. They paid the market rate, which could be high if a worker was very productive and low if he was not. "Many of the English working classes have not been properly fed, and scarcely any of them have been properly educated." Low productivity was the cause of "low wages of a large part of the English people and of the actual pauperism of no inconsiderable number." And although Marshall did not deny that "there is any form of land nationalization which, on the whole, would benefit," he argued that "there is none that contains a magic and sudden remedy for poverty. We must be content to look for a less sensational cure." [123]

That cure, Marshall said, was to raise productivity. One way was to:

> educate (in the broadest sense) the unskilled and inefficient workers out of existence. On the other hand—and this sentence is the kernel of all I have to say about poverty—if the numbers of unskilled laborers were to diminish sufficiently, then those who did unskilled work would have to be paid good wages. If total production has not increased, these extra wages would have to be paid out of the shares of capital and of higher kinds of labor . . . But if the diminution of unskilled labor is brought about by the increasing efficiency of labor, it will increase production, and there will be a larger fund to be divided up.

He did not object to unions or even to some fairly radical proposals for land reform or progressive taxation. He merely noted that none of these could produce "more bread and butter." This required "competition," time, and the cooperation of all parts of society, government, and the poor themselves.[124]

He accused George of promoting a quack cure. The problem wasn't just that "Mr. George said, 'If you want to get rich, take land,'" but that it would divert from education and training, hard work, and thrift. George's scheme would yield "less than a penny in the shilling on their income . . . For the sake of this, Mr. George is willing to pour contempt on all the plans by which workingmen have striven to benefit themselves."[125]

When Marshall's *Principles of Economics* finally appeared in 1890, it breathed new life into a faltering discipline. It established him as its intellectual leader and the authority to whom governments turned for advice.

Principles embodied Marshall's rejection of Socialism, embrace of the system of private property and competition, and optimism about the improvability of man and his circumstances. The book portrayed economics not as a dogma but as "an apparatus of the mind." As Dickens hoped, Marshall had managed, while placing the discipline on a more sound scientific footing, to humanize economics by injecting "a little human bloom . . . and a little human warmth."

But the chief insight reflected the lesson he learned in America. Under a system of private property and competition, business firms are under constant pressure to achieve more with the same or fewer resources. From society's standpoint, the corporation's function is to raise productivity and, hence, living standards.

Of all social institutions, the business firm was more central, enjoyed a higher status, and did more to shape the American mind and civilization than elsewhere. The company was not only the principal creator of wealth in America but also the most important agent of social change and the biggest magnet for talented individuals. It made Dickens's depictions of businessmen as cretins or predators, workers as zombies, and successful manufacture as rigid repetition look ridiculous. The undisputed fact that American productive power was growing at an unimaginably rapid rate meant that businesses must be doing more, at least in the aggregate, than exploiting Peter to line Paul's pockets or merely repeating the same operations from one year to the next. On his visits to factories, Marshall was especially struck by managers' constant search for small improvements and workers' equally constant search for better opportunities and useful skills. Both seemed obsessed with making the most of the resources at their command.

Naturally, Marshall recognized that companies also exist to generate profits for owners, managerial salaries for executives, and wages for workers. Adam Smith had pointed out that to maximize their own income in the face of competition, firms had to benefit consumers by producing as much and as cheaply as possible. But Marshall introduced the element of time into his analysis. Over time, firms could remain profitable and continue to exist only if they became more and more productive. Survival in the face of competition not only implied incessant adaptation. Competition for the most productive workers meant that, over time, firms had to share gains from productivity improvements.

This is precisely what Mill and the other founders of political economy had denied. They had maintained that advances in productivity were of little or no benefit to the working classes. In their imaginary firms, productivity might grow by leaps and bounds, but wages never rose for long above some physiological maximum. Working conditions, if anything, worsened over time. Marshall saw not only that this was not so in fact, but also that it could not be so. Competition for labor forced owners to share the benefits of efficiency and quality improvements with workers, first as wage earners, then as consumers. The evidence confirmed that Marshall was right. The share of wages in the gross domestic product—the nation's annual income from wages, profits, interest, and proprietors' income—was rising, not falling, and so were the levels of wages and working-class consumption—as they had been in most years since 1848, when The Communist Manifesto and Mill's Principles of Political Economy appeared.

Chapter III

Miss Potter's Profession:
Webb and the Housekeeping State

> She yearned for something by which her life might be filled with
> action at once rational and ardent; and since the time was gone
> by for guiding visions . . . what lamp was there but knowledge?
>
> —*George Eliot*, Middlemarch[1]

Every year in March, the "upper ten thousand" descended on London like
a vast flock of extravagantly plumed and exotic migratory birds.[2] During
the three or four months of the London "season," Britain's elite devoted
itself to an elaborate mating ritual. Mornings were spent riding along Rotten Row or the Ladies' Mile in Hyde Park. Afternoons were for repairing
to Parliament or to clubs for the males of the species, shopping and paying
social calls for the wives and daughters. In the evenings everyone met at
operas, dinner parties, and balls that provided opportunities for magnificent displays. Every few days, an obligatory race, regatta, cricket match, or
gallery opening introduced a slight variation to the schedule.

As with so much else in Victorian high society, this frenetic and seemingly frivolous pursuit of pleasure was serious business: during the season,
which began when Parliament resumed its session, London became the
epicenter of the global marriage market. Wealthy parents thought of giving a daughter two or three London seasons in the same way as sending
a son to Oxford or Cambridge. The expense and effort of participating
in this extraordinarily intricate mating dance were certainly comparable.

If the family had no permanent "town" house, an imposing mansion had to be found at a fashionable address. A vast quantity of expensive items had to be purchased and conveyed, what with "the stable of horses and carriages . . . , the elaborate stock of garments . . . , [and] all the commissariat and paraphernalia for dinners, dances, picnics and weekend parties" considered de rigueur. Needless to say, socializing on so ambitious a scale demanded an executive to oversee "extensive [plans], a large number of employees and innumerable decisions"—in other words, the lady of the household.[3]

These were the reflections that occupied Beatrice Ellen Potter, Bo or Bea to her family, the eighth of nine daughters of a rich railway tycoon from Gloucester named Richard Potter. The carriage she was sharing with her father on a raw February afternoon in 1883 rolled to a stop in front of an imposing terrace of tall, cream-colored Italianate villas. The slender young woman with an air of command surveyed number 47 Princes Gate coolly. It was meant to serve as the social headquarters of the sprawling Potter clan, which included her six married sisters and their large families, for the season. The five-story mansion had a sumptuous façade with Ionic columns, Corinthian pilasters, tall windows, and swags of fruit and flowers, and it faced Hyde Park. At the back, visible through French doors, was an expanse of terraced lawn furnished with classical statues and enormous pots with masses of scarlet pelargoniums tumbling over their sides. The houses on either side of theirs were similarly grand. Her father had chosen Princes Gate precisely so they would be flanked by neighbors as wealthy and powerful as he. Junius Morgan, the American banker, leased number 13. Joseph Chamberlain, a Manchester industrialist turned Liberal politician and the father of Neville Chamberlain, had taken number 40 for the season. It was a perfect setting for Potter's brilliant daughter.

At twenty-five, Beatrice was the veteran of more than half a dozen London seasons but had never been in love. Until now, her duties had consisted of enjoying herself at some fifty balls, sixty parties, thirty dinners, and twenty-five breakfasts before society packed up and retreated to the country in July.[4] She'd had nothing whatsoever to do with "all that elaborate machinery"[5] that was required backstage. This year would be different. Beatrice had been the only one of the Potter sisters, except for thirteen-year-old Rosie, who was still living at home in Gloucester when their mother died the previous spring. Suddenly she was promoted to lady of her father's house.

Before leaving Gloucester, Beatrice had made a solemn vow "to give myself up to Society, and make it my aim to succeed therein."[6] By "succeed," she meant marry an important man, as each of her older sisters had done, although her use of "give myself up" suggests that the price of success was self-immolation. The latest to do so had been her favorite sister, Kate, who had waited until the advanced age of thirty-one to marry a prominent Liberal economist and politician, Leonard Courtney, currently serving as Treasury secretary. Her father did not doubt that Bo would follow suit. Besides beauty, breeding, and a large fortune, she had the gift of commanding attention. Her long, graceful neck, fiercely intelligent eyes, and glossy black hair made people seeing her for the first time across a crowded room think of a beautiful, slightly dangerous black swan. Men were charmed by her, especially when they realized that she refused to take them seriously.

For a while after the Potters' arrival, all was chaos and confusion. More retainers, extra horses, and additional carriages appeared. When the servants finally withdrew and her father had gotten some supper, Beatrice went upstairs and found the bedroom in the back of the mansion that she had determined would be hers. Now she could think of something besides seating plans and menus—namely, the books that she had brought to read; the things she meant to learn. Beatrice saw nothing inherently contradictory in her various desires and duties. After all, a happily married woman sat on the throne, and George Eliot reigned as the most successful writer of the day. When Beatrice was eighteen, she had spent more time studying Eastern religions than preparing for her "coming out."

Her bedroom window overlooked the Victoria and Albert Museum. It suddenly struck her that the great monument to human ingenuity stood in the very center of London yet managed to remain wonderfully "undisturbed by the rushing life of the great city."[7] Beatrice wondered whether she might do the same, maintaining a Buddhist-like detachment in crowded drawing rooms and theaters. Might she not she fulfill society's expectations while still cultivating the "thoughtful" part of her life, the part that drove her to constantly to ask herself, "How am I to live and for what object?"[8]

The question of her destiny had preoccupied Beatrice since her fifteenth birthday. Her mother and sisters had always regarded her obsession as unhealthy. Wasn't it enough to simply be "one of the fashionable Miss Potters who live in grand houses and beautiful gardens and marry enormously wealthy men?"[9] Had Beatrice been the heroine of a Victorian novel, its author would have felt obliged to offer some justification for making the question of her destiny the "center of interest." In *The Portrait of a Lady,* published in 1881, Henry James had done exactly that: "Millions of presumptuous girls, intelligent or not intelligent, daily affront their destiny, and what is it open to their destiny to be, at the most, that we should make an ado about it?" James asked in his preface.[10] Before middle-class women had viable alternatives to early marriage and motherhood, and the Married Women's Property Act of 1882 gave them the right to their own incomes, the recurring question in *The Portrait of a Lady*—"Well, what will she do?"—could hardly have merited a reader's interest.

"You are young, pretty, rich, clever, what more do you want?" Beatrice's cousin Margaret Harkness, the novelist daughter of a poor country parson, had asked with a trace of exasperation when they were at school together. "Why cannot you be satisfied?"[11] Like James's heroine Isabel Archer, Beatrice had been brought up with an unusual freedom to travel, read, form friendships, and satisfy her "great desire for knowledge" and "immense curiosity about life." Beatrice preferred the company of men and took it for granted that most would fall under her spell, but like Isabel she had no desire to "begin life by marrying."[12] She was as interested in winning recognition for her intellectual achievements as for her feminine charms. Each passing year made her longing for a "real aim and occupation" more urgent.[13] She was conscious of "a special mission" and believed with all her heart that she was meant to live "a life with some result."[14] Like

Dorothea in *Middlemarch,* Beatrice yearned for principles, "something by which her life might be filled with action at once rational and ardent."[15]

Beatrice's identity was shaped by having been born into Britain's "new ruling class,"[16] and her mind by having been "brought up in the midst of capitalist speculation" and "the restless spirit of big enterprise."[17] As the historian Barbara Caine notes, Beatrice distinguished her class not by wealth but by the fact that it was "a class of persons who habitually gave orders, but who seldom, if ever, executed the orders of other people."[18] Her grandfathers were both self-made men. Her father had lost the lion's share of his inheritance in the crash of 1848, only to quickly recoup his losses by supplying tents to the French army during the Crimean War. By the time Beatrice was born in 1858, Richard Potter had amassed a third fortune from timber and railways and had become a director (and future chairman of the board) of the Great Western Railway. More entrepreneur and speculator than hands-on manager, Potter once toyed with a scheme to build a waterway to rival the Suez Canal. His business interests were scattered from Turkey to Canada, and he and his family were constantly traveling. The Potters' Gloucester estate, Standish, as grand and impersonal as a hotel, was filled with an ever-changing contingent of visiting relatives, guests, employees, and hangers-on.

Though Richard Potter began to vote Conservative in middle age, he was never a stereotypical Tory plutocrat. His father, a wholesaler in the cotton industry, was for a time a Radical member of Parliament and had helped found the *Manchester Guardian*[19] ("our organ," as Beatrice used to say).[20] Intellectually engaged, broad-minded, and convivial, he counted scientists, philosophers, and journalists among his closest friends. Herbert Spencer, the most lionized intellectual in England in the 1860s and 1870s and a former railway engineer and *Economist* editorialist, called Richard Potter "the most lovable being I have yet seen."[21] Even the latter's cheerful indifference to Spencer's philosophical interests could not squelch Spencer's lifelong adoration.

It is almost axiomatic that behind every extraordinary woman there is a remarkable father. Potter encouraged Beatrice and her sisters to read and gave them free run of his large library. He made no effort to restrict their discussion or friendships. He enjoyed their company so much that he rarely took a business trip without one or another as a companion. Beatrice claimed that "he was the only man I ever knew who genuinely believed that women were superior to men, and acted as if he did." [22] She gave him credit for her own "audacity and pluck and my familiarity with the risks and chances of big enterprises." [23]

In some ways, Laurencina Potter was even more unusual than her husband, bearing even less resemblance to the plump and placid mothers that populate Trollope novels than her husband did to the stereotypical businessmen. When Spencer met the Potters for the first time shortly after their marriage, he thought they were "the most admirable pair I have ever seen." [24] As he got to know them better, he was surprised to learn that Laurencina's perfectly feminine, graceful, and refined personage hid "so independent a character." [25] In contrast to her easygoing husband, Laurencina was cerebral, puritanical, and discontented. Born Heyworth, she came from a family of liberal Liverpool merchants who educated her no differently from her brothers; that is, trained her in mathematics, languages, and political economy. As a young woman, she became a local celebrity and the subject of newspaper articles as a result of the zeal with which she canvassed against the Corn Laws. Decades later, Beatrice was used to seeing pamphlets on economic issues appear on her dressing table.

Laurencina was a very unhappy woman. The cause of her frustration was not hard for her daughter to divine. She had envisioned a married life of "close intellectual comradeship with my father, possibly of intellectual achievement, surrounded by distinguished friends." [26] Instead, for the first two decades after her marriage, she was almost always pregnant or nursing an infant and banished to the company of women and children while her husband traveled on business and dined with writers and intellectuals. Her real ambition had been to write novels, and she did publish one, *Laura Gay,* before the demands of family life overwhelmed her.

When Laurencina's ninth child and only son, Dickie, was born, she devoted herself completely to him. But when the boy died of scarlet fever at age two, she became severely depressed and withdrew from her other children. Beatrice, who was seven at the time, recalled her mother as "a remote personage discussing business with my father or poring over books in her boudoir." As a consequence of her mother's coldness, Beatrice came to believe that "I was not made to be loved; there must be something repulsive about my character." Moody, self-dramatizing, and inclined to fib and exaggerate, she had also inherited the Heyworths' tendency to Weltschmerz and suicide. Two of Laurencina's relatives had died by their own hand. "My childhood was not on the whole a happy one," reflected Beatrice as an adult. "Ill-health and starved affection and the mental disorders which spring from these, ill temper and resentment, marred it . . . Its *loneliness* was absolute."[27] Beatrice herself had toyed with bottles of chloroform even as a child.

Rebuffed by her mother, a biographer says, Beatrice sought affection "below stairs" among the servants who helped keep the Potter household running. She and her older sisters were especially close to Martha Jackson, or Dada as they called her, who took care of the children. Dada, Beatrice learned much later, was actually a relative from a branch of her mother's family who were poor but respectable Lancashire hand-loom weavers. Caine credits Dada with planting in Beatrice the notion of original sin that gave her the determination to do good and the identification she felt all her life with the "respectable" working poor. But it was Laurencina's example that inspired her to write. On her fifteenth birthday, Beatrice began a journal that she kept until her death. "Sometimes I feel as if I must write, as if I must pour my poor crooked thoughts into somebody's heart, even if it be into my own."[28]

Among the intellectuals who frequented the Potter house were the biologist Thomas Huxley; Sir Francis Galton, a cousin of Charles Darwin's; and other proponents of the new "scientific" point of view that was undermining traditional beliefs. By the time Beatrice was in her teens, Spencer, who shared the Potters' background as dissenting Protestants, had become Laurencina's closest confidant and the dominant intellectual influence in the household.

Spencer, who coined the term "survival of the fittest," was a bigger celebrity in the 1860s than Charles Darwin. His notion that social institutions, like animal or plant species, were evolving—and therefore could be observed, classified, and analyzed like plants or animals—had captured the public's imagination. One of the earliest exponents of evolution, Spencer was a radical individualist who opposed slavery and supported women's suffrage. His antipathy toward government regulation and high taxes appealed to the upwardly mobile middle and lower-middle classes. His popularity was further bolstered by his refusal to rule out absolutely the existence of God.

Fame, however, had not agreed with Spencer. In uncertain health and prone to hypochondria, he had grown increasingly reclusive and eccentric with age. When not at his club or alone in his rooms, he turned to the company of the Potters and their children. A frequent guest at the Potters' Gloucester estate, he delighted in liberating the Potter sisters from their governesses while chanting, "Submission *not* desirable."[29] Often he would lead them off to gather specimens to illustrate one or another of his ideas about evolution. In the summers, when the Potters were at their Cotswold retreat, he would lead the way through beech groves and old pear orchards, dressed in white linen from head to toe, carrying a parasol. Trailing behind would be a "very pretty and original group"[30] of tall, slender girls with boyishly short black hair, dressed in pale muslins and carrying buckets and nets. From time to time, the party stopped to dig for fossils. The old railway cuts or limestone quarries of Gloucester once lay under shallow, warm seas and thousands of years later yielded a choice collection of ammonites, crinoids, trilobites, and echinoids. The girls did not take their hyperrational friend too seriously. "Are we descended from monkeys, Mr. Spencer?" they would chorus amid giggles. His unvarying reply—"About 99 percent of humanity have descended and one percent have ascended!"—elicited more peals of mirth as well as, occasionally, volleys of decaying beech leaves aimed at the philosopher's "remarkable headpiece."[31]

The most bookish and moodiest of the sisters, Beatrice developed a lasting fascination with the workings of Spencer's remarkable mind. Spencer encouraged Beatrice by telling her that she was a "born metaphysician," comparing her to his idol, George Eliot, drawing up reading lists, and urging her to pursue her intellectual ambitions. Without his support, Beatrice might have submitted to the kind of life that convention—and at times her own heart—demanded.

Beatrice's formal education was shockingly skimpy. Like that of many upper-class young women, it was limited to a few months at a fancy finishing school, in part because of her own frequent illnesses, both imaginary and real, and in part because even Richard Potter, liberal as he was by the standards of the time, never thought of sending her to university. She was therefore largely educated at home—that is to say, self-taught and free to read even books that had been banned from public libraries. "I am, as Mother says, too young, too uneducated, and worst of all, too frivolous to be a companion to her," she wrote in her diary. "But, however, I must take courage, and try to change." [32] Penny-pinching in most things, Laurencina was openhanded when it came to buying newspapers and magazines. Beatrice plunged into religion, philosophy, and psychology, her mother's interests. Her schoolgirl reading included George Eliot and the fashionable French philosopher and pioneering sociologist Auguste Comte.

Because Beatrice was given unlimited access to her father's library and her mother's journals, she was exposed in a way few girls were to the religious and scientific controversies that dominated the late Victorian era. "We lived, indeed, in a perpetual state of ferment, receiving and questioning all contemporary hypotheses as to the duty and destiny of man in this world and the next," she recalled. By the time Beatrice was eighteen and about to come out, she had substituted for the old Anglican faith Spencer's new doctrine of "harmony and progress." She had also embraced her mentor's libertarian political creed and his ideal of the "scientific investigator." The image of the latter aroused her "domineering curiosity in the nature of things" and "hope for a 'bird's eye view' of mankind" as well as her secret ambition to write "a book that would be read." [33]

After three weeks at Princes Gate, Beatrice was suffering from the "rival pulls on time and energy." [34] After a particularly tedious dinner party, she fumed that "Ladies are so expressionless." [35] She no longer understood why "intelligent women wish to marry into the set where this is the social regime."[36] She poured her discontents into her diary: "I feel like a caged animal, bound up by the luxury, comfort, and respectability of my position."[37]

Beatrice longed for work as well as love, but she was beginning to wonder whether her chances of having it all were any better than poor Laurencina's. When Isabel Archer insisted that "there are other things a woman can do," she was thinking, presumably, of the small but growing ranks of self-supporting female professionals who could befriend whomever they pleased, talk about whatever they liked, live in lodgings, and travel on their own.

But such women gave up a great deal, Beatrice realized upon reflection. When she encountered the daughter of the notorious Karl Marx in the refreshment room of the British Museum, Eleanor Marx was "dressed in a slovenly picturesque way with curly black hair flying about in all directions!" Beatrice was taken by Eleanor's intellectual self-confidence and romantic appearance but repelled by the latter's bohemian lifestyle. "Unfortunately one cannot mix with human beings without becoming more or less *connected* with them," she told herself.[38] She adored her cousin Margaret Harkness, the future author of *In Darkest London, A City Girl,* and other social novels. Maggie lived on her own in a seedy one-room flat in Bloomsbury and had tried teaching, nursing, and acting before discovering her talents as a writer. Her family was horrified, and Maggie had been forced to break off all contact with them, something Beatrice could no more imagine than she could picture immigrating to America. She wished that she could be more contented. "Why should I, wretched little frog, try to puff myself into a professional? If I could rid myself of that mischievous desire to achieve . . ."[39]

Once again Spencer came to the rescue by suggesting that Beatrice take her older sister's place as a volunteer rent collector in the East End. She could prepare for a career of social investigation while continuing her private studies. Like Alfred Marshall a generation before, Beatrice found herself drawn to London. She went off to a meeting of the Charity Organization Society, a private group dedicated to "scientific" or evidence-based charity and the gospel of self-help. "People should support themselves by their own earnings and efforts and . . . depend as little as possible on the state."[40] Women had traditionally been responsible for visiting the poor, but by the 1880s social work was becoming a respectable profession for spinsters and married women without children. The attractions were manifold. Beatrice observed: "It is distinctly advantageous to us to go amongst the poor . . . We can get from them an experience of life which is novel and interesting; the study of their lives and surrounding gives us the facts whether with we can attempt to solve the social problems."[41] Shortly afterward she thought, "If I could only devote my life to it . . ."[42] Yet, as of a few months earlier, Beatrice had made only two or three visits to the Katherine Houses in Whitechapel. "I can't get the training that I want without neglecting my duty," she sighed.[43]

One night that same month, Beatrice lay awake until dawn, too excited to sleep. Her partner at a neighbor's dinner party had been Joseph Chamberlain, the most important politician in England and the most commanding and charismatic man she had ever met.

Chamberlain was twenty-two years older than Beatrice and twice widowed, but he radiated youthful vigor and enthusiasm. Powerfully built with thick hair, a piercing gaze, and a curiously seductive voice, he was a natural leader. He had made a large fortune as a manufacturer of screws and bolts before moving into politics as the reform-minded mayor of Birmingham. For four years, he "parked, paved, assized, marketed, Gas and Watered and *improved*"[44] the grimy factory town into a model metropolis. After spending several years rebuilding the Liberal Party's crumbling political machine, he was rewarded with a cabinet post.

By the time Beatrice met Chamberlain, he had become the bad boy of English politics. His studied elegance—contrived with a monocle, a bespoke suit, and a fresh orchid on his lapel—hardly fit his rabble-rouser image. But in the stormy debates of that year, Chamberlain had focused voters' attention on the twin issues of poverty and voting rights. He had used his cabinet post to campaign for universal male suffrage, cheaper housing, and free land for farm laborers. He infuriated Conservatives by inviting the party's leader, Lord Salisbury, to visit Birmingham—only to serve as the keynote speaker at a rally protesting his presence. His rivals called him the "English Robespierre" and accused him of fomenting class hatred. Queen Victoria demanded that Chamberlain apologize after insulting the royal family at a working-class demonstration. Herbert Spencer told Beatrice that Chamberlain was "a man who may mean well, but who does and will do, an incalculable amount of mischief."[45]

As a disciple of Spencer's, Beatrice disapproved of virtually everything Chamberlain stood for, especially his populist appeals to voters' emotions. Nonetheless, he excited her. "I do, and don't, like him," she wrote in her diary. Sensing danger, she warned herself sternly that "talking to 'clever men' in society is a snare and delusion . . . Much better read their books."[46] She did not, however, follow her own advice.

Given that the Potters and Chamberlain were neighbors at Princes Gate, it was inevitable that the controversial Liberal politician and the fashionable, if slightly unconventional, Miss Potter should be constantly thrown together. The second time they met was that July, at Herbert Spencer's annual picnic. After spending the entire evening in conversation with Chamberlain, Beatrice admitted, "His personality interested me."[47] A couple of weeks later, she found herself seated between Chamberlain and an aristocrat with vast estates. "Whig peer talked of his own possessions, Chamberlain *passionately* of getting hold of other people's—for the masses," she joked. Though she found his political views distasteful, she was captivated by his "intellectual passions" and "any amount of *purpose.*" Beatrice thought to herself: "How I should like to study that man!"[48]

Beatrice was fooling herself. The social investigator and detached observer had already lost her footing and slipped into the "whirlpool" of emotions to which she was irresistibly drawn but that she could neither comprehend nor control. She agonized over whether or not she would be happy as Chamberlain's wife. Used to charming the men around her, she was unsatisfied by easy conquests. Starved for affection as a child, she longed to capture the attention of a man who was focused not on her but on some important pursuit. Chamberlain, who wanted to be prime minister, demanded blind loyalty from followers and family alike, and seduced crowds the way other men seduced women. He was the most powerful personality Beatrice had ever encountered. Might he not relish a strong mate?

She tried to analyze his peculiar fascination for her: "The commonplaces of love have always bored me," she wrote in her diary.

> But Joseph Chamberlain with gloom and seriousness, with his absence of any gallantry or faculty for saying pretty nothings, the simple way in which he assumes, almost asserts, that you stand on a level far beneath him and that all that concerns you is trivial; that you yourself are without importance in the world except in so far as you might be related to him; this sort of courtship (if it is to be called courtship) fascinates, at least, my imagination.[49]

Beatrice half expected Chamberlain to declare himself before the end of the London season, but no offer of marriage was forthcoming. Disappointed, Beatrice returned to Standish, where she "dreamt of future achievement or perchance of—love."[50] In September, Chamberlain's sister, Clara, invited her to visit Chamberlain's London house. Again Beatrice assumed that Chamberlain would propose. "Coming from such honest surroundings he surely *must* be straight in intention," she told herself.[51] Again, he did not, even though his intentions had become a topic of discussion within the Potter family. Beatrice tried to lower her own and her sisters' expectations: "If, as Miss Chamberlain says, the Right Honorable gentleman takes 'a very conventional view of women,' I may be saved all temptation by my unconventionality. I certainly shall not hide it."[52]

In October, while Beatrice was at Standish obsessing over Chamberlain, the Liberal *Pall Mall Gazette* excerpted a first-person pamphlet about London's East End by a Congregational minister.[53] The series exposed deplorable housing conditions in gruesome detail that scandalized and galvanized the middle classes. Like Henry Mayhew's eyewitness accounts of poverty in the 1840s and 1850s, "The Bitter Cry of Outcast London" chronicled crowding, homelessness, low wages, disease, dirt, and starvation. But as Gertrude Himmelfarb points out, its shock value depended even more on its hints of promiscuity, prostitution, and incest:

> Immorality is but the natural outcome of conditions like these. . . . Ask if the men and women living together in these rookeries are married, and your simplicity will cause a smile. Nobody knows. Nobody cares. . . . Incest is common; and no form of vice and sensuality causes surprise or attracts attention.[54]

The immediate effect of the sensational expose was to goad Lord Salisbury, the prime minister, and Joseph Chamberlain into a debate over the cause of the crisis and the government's response. The Tory leader and major landowner in the East End blamed London's infrastructure boom for overcrowding, while Chamberlain placed the blame on urban property owners, whom he wanted to tax to pay for worker housing. Significantly, both the Tory and the Radical assumed that the responsibility for the housing crisis belonged to the government.

Beatrice dismissed the *Pall Mall* series as "shallow and sensational" and joined Spencer in regretting its political impact.[55] She recognized, however, that its first-person testimony and personal observations accounted for the extraordinary reception. She had, as she reminded herself, been led into tenements not by the spirit of charity but by the spirit of inquiry. The stupendous reaction to "The Bitter Cry"—and Spencer's hope that someone who shared his views would produce an effective rebuttal—made her eager to test her own powers of social diagnosis.

Beatrice decided to begin on relatively familiar ground by visiting her mother's poor relations in Bacup, in the heart of the cotton country. These included her beloved Dada, who had married the Potters' butler. It is a measure of the independence Beatrice enjoyed that she could undertake such a project. To avoid embarrassing her family and rendering her interviewees speechless, she went to Lancashire not as one of the "grand Potters" but merely as "Miss Jones." After a week, she wrote to her father, "Certainly the way to see industrial life is to live amongst the workers."[56]

She found what she had prepared herself to find: "Mere philanthropists are apt to overlook the existence of an independent working class and when they talk sentimentally of 'the people' they mean really the 'ne'er do weels.'" [57] She decided to write a piece about the independent poor. When she saw Spencer at Christmas, he urged her to publish her Bacup experiences. Actual observation of the "working man in his normal state" was the best antidote to "the pernicious tendency of political activity" on the part of Tories as well as Liberals toward higher taxes and more government provision.[58] Spencer promised to talk to the editor of the magazine *The Nineteenth Century.* Naturally, Beatrice, was extremely gratified, but she was also secretly amused that "the very embodiment of the 'pernicious tendency'" had not only captured his protégé's heart but was also about to invade the Potter family circle.[59]

Beatrice had invited Chamberlain and his two children to Standish at the New Year. She could see no way to resolve her divided feelings without a face-to-face meeting and she was sure that he must feel the same way: "My tortured state cannot long endure," she wrote in her diary. "The 'to be or not to be' will soon be settled." [60] Instead, the visit proved horribly awkward. The more Beatrice resisted Chamberlain's political views, the more forcefully he reiterated them, leading him to complain after one heated match that he felt as if he had been giving a speech. "I felt his curious scrutinizing eyes noting each movement as if he were anxious to ascertain whether I yielded to his absolute supremacy," Beatrice noted. When Chamberlain told her that he merely desired "intelligent sympathy" from women, she snidely accused him in her mind of really wanting "intelligent servility." Once again, he left without proposing.[61]

"If you believe in Herbert Spencer, you won't believe in me," Chamberlain had flung at Beatrice during their last exchange.[62] If he hoped to convert her, he was mistaken.

When Beatrice was a girl, her father used to tease Spencer for his habit of "walking against the tide of churchgoers" in the village near the Potter estate. "Won't work, my dear Spencer, won't work," Richard Potter would murmur.[63] But for two decades or more, Spencer had an entire generation of thinking men and women following his lead. His *Social Statics,* published within three years of the revolutions of 1848 throughout Europe, had celebrated the triumph of new economic and political freedoms over aristocratic privilege and made minimal government and maximum liberty the creed of middle-class progressives. Alfred Marshall imbibed more evolutionary theory from Spencer than from Darwin. Karl Marx sent Spencer a signed copy of the second edition of *Das Kapital* in the hopes that the philosopher's endorsement would boost its sales.[64]

By the early 1880s, however, Spencer was again walking against the tide. His latest book, *The Man Versus the State,* was a sweeping indictment of the steady growth of government regulation and taxation:

> Dictatorial measures, rapidly multiplied, have tended continually to narrow the liberties of individuals; and have done this in a double way. Regulations have been made in yearly-growing numbers, restraining the citizen in directions where his actions were previously unchecked, and compelling actions which previously he might perform or not as he liked; and at the same time heavier public burdens, chiefly local, have further restricted his freedom, by lessening that portion of his earnings which he can spend as he pleases, and augmenting the portion taken from him to be spent as public agents please.[65]

His brief for laissez-faire struck the reading public as a last-ditch defense of an outmoded, reactionary, and increasingly irrelevant doctrine. As Himmelfarb explains, not only were most thinking Victorians moving away from, or at least questioning, laissez-faire, but many now regretted that they had ever embraced it. She cites Arnold Toynbee, the Oxford economic historian, who apologized to a working-class audience: "We—the middle classes, I mean not merely the very rich—we have neglected you; instead of justice we have offered you charity."[66]

When Spencer's book appeared in 1884, he and Beatrice were closer than ever, spending several hours a day together. "I understand the working of Herbert Spencer's reason; but I do not understand the reason of Mr. Chamberlain's passion," she admitted.[67] She sent her signed copy of

The Man Versus the State to the mistress of Girton College at Cambridge with a note that shows that she remained the most ardent of Spencer's disciples. Referring to relief for the jobless, public schools, safety regulations, and other instances of large-scale "state intervention," she wrote, "I object to these gigantic experiments . . . which flavor of inadequately—thought-out theories—the most dangerous of all social poisons . . . the crude prescriptions of social quacks."[68]

Yet, she was ambivalent. Chamberlain had forced her to recognize that "social questions are the vital questions of today. They take the place of religion."[69] So while she was not prepared to embrace the new "time spirit" overnight, she was not ready to dismiss it out of hand, much less give up its virile and forceful proponent.[70]

When Chamberlain's sister invited her to visit Highbury, his massive new mansion in Birmingham, Beatrice went at once, assuming that the invitation originated with her chosen lover. But as soon as she arrived she was struck by the incompatibility of their tastes. She found nothing to praise about the "elaborately built red-brick house with numberless bow windows" and could barely repress a shudder when confronted with its vulgar interior of "elaborately-carved marble arches, its satin paper, rich hangings and choice watercolors . . . forlornly grand. No books, no work, no music, not even a harmless antimacassar, to relieve the oppressive richness of the satin-covered furniture."

On Beatrice's first day there, John Bright, an elder statesman of the Liberal Party, regaled her with reminiscences of her mother's brilliance as "girl-hostess" to the teetotalers and Anti–Corn Law League enthusiasts who visited the Heyworth house forty years earlier, recalling Laurencina's political courage during the anti–Corn Law campaign. The old man's expression of admiration for her mother's political faculty and activism made Chamberlain's insistence that the women in his house have no independent opinions seem even more despotic. But Chamberlain's egotism attracted Beatrice. That evening at the Birmingham Town Hall, she watched him seduce a crowd of thousands and dominate it completely. Beatrice mocked the constituency as uneducated and unquestioning, hypnotized by Chamberlain's passionate speaking and not his ideas, but seeing "the submission of the whole town to his autocratic rule," she admitted that her own surrender was inevitable. Chamberlain would rule the same way at home and even her own feelings would betray her. ("When feeling becomes strong, as it would do with me in marriage, it would mean the absolute subordination of the reason to it.") Even knowing that Chamberlain would make her miserable, Beatrice was caught. "His personality absorbs all my thought," she wrote in her diary.

The next morning, Chamberlain made a great point of taking Beatrice on a tour of his vast new "orchid house." Beatrice declared that the only flowers *she* loved were wildflowers, and feigned surprise when Chamberlain appeared annoyed. That evening, Beatrice thought she detected in his looks and manner an "intense desire that I should *think and feel like him*" and "jealousy of other influences." She took this to mean that his "susceptibility" to her was growing.[71]

In January 1885, Chamberlain was launching the most radical and flamboyant campaign of his career. He enraged his fellow Liberals by warning his working-class constituents that the franchise wouldn't lead to real democracy unless they organized themselves politically. He scandalized Conservatives by ratcheting up the rhetoric of class warfare with the famous phrase "I ask what ransom property will pay for the security which it enjoys?"[72] Having administered Birmingham on the bold principle of "high rates and a healthy city," Chamberlain took advantage of his cabinet position to demand universal male suffrage, free secular education, and "three acres and a cow" for those who preferred individual production on the land to work for wages in the mine or the factory. These were to be paid for by higher taxes on land, profits, and inheritances. Once again Beatrice went to Birmingham and sat in the gallery while he delivered a fiery peroration, and the next day, she again experienced the humiliation of rejection. He did not propose.

Beatrice's obsessive, conflicted passion continued to torment her. She despised herself for being infatuated with a domineering man, but also for failing to conquer him. She had dared to hope for a life combining love and intellectual achievement. She had, at different times, been ready to sacrifice one for the sake of the other. Now it seemed to her that she had been deluded about her own potential from the start. "I see clearly that my intellectual faculty is only mirage, that I have no special mission" and "I have loved and lost; but possibly by my own willful mishandling, possibly also for my own happiness; but still lost."[73]

In her dejected state, she marveled that she had ever aspired to win an extraordinary man like Chamberlain and tortured herself with what might have been: "If I had from the first believed in that purpose, if the influence which formed me and the natural tendency of my character, if they had been different, I might have been his helpmate. It would not have been a happy life; it might have been a noble one."[74] On the first of August, she made her will: "In case of my death I should wish that all these diary-books, after being read (if he shall care to) by Father, should be sent to Carrie Darling [a friend]. Beatrice Potter."[75]

Somehow, she recovered from the blow. By the general election in early November 1885, suicide no longer dominated her thoughts, and she could feel a bit of her energy returning. As she watched her father set off for the polls to cast his vote, she was once again plotting her career as social investigator. That is when fate landed her another blow that threatened to bring independent action "to a sudden and disastrous end."[76] Richard Potter was brought back to Standish from the polls, having suffered a devastating paralytic stroke that did not, however, kill him.

As always, Beatrice poured her despair into her diary. "Companionizing a failing mind—a life without physical or mental activity—no work. Good God, how awful."[77] On New Year's Day, she drafted another will, begging the reader to destroy her diary after her death. "If Death comes it will be welcome," she wrote bitterly. "The position of an unmarried daughter at home is an unhappy one even for a strong woman: it is an impossible one for a weak one."[78]

Now, her old obsession with how she was to live, what purpose she would achieve, and whom she would love seemed like the purest hubris. "I am never at peace with myself now," she wrote in early February 1886. "The whole of my past life looks like an irretrievable blunder, the last two years like a nightmare! . . . When will pain cease?"[79]

• • •

The answer came a few days later in the form of a mighty roar that seemed to come from society's hidden depths. By noon on Monday, February 8, a crowd of ten thousand had braved fog and frost to assemble in Trafalgar Square. Some 2,500 police ringed the square's perimeter. They estimated that two-thirds of the crowd consisted of unemployed workmen, the rest radicals of every conceivable stripe. A Socialist speaker, chased off the base of Admiral Nelson's statue earlier that morning, clambered back up unhindered by the authorities. He waved a red flag defiantly and fired up the crowd with denunciations of "the authors of the present distress in England."[80] On behalf of his listeners, he demanded that Parliament provide public works jobs for "the tens of thousands of deserving men who were out of work through no fault of their own."[81] The audience cheered and swelled throughout the afternoon until it had grown fivefold.

The rally ended peaceably, but then the demonstrators began pouring into the main streets of the West End—Oxford Street, St. James Street, the Pall Mall—"cursing the authorities, attacking shops, sacking saloons, getting drunk, and smashing windows." The police were not only caught off guard but grossly outnumbered. For three hours or more, a "hooting howling mob" ruled the West End. Hundreds of shops were looted, anyone who looked like a foreigner was beaten, a Lord Limerick was pinned to the railings of his club, and carriages in Hyde Park were overturned and robbed. In addition, all street traffic in central London came to a standstill, Charing Cross Station was completely paralyzed, and by nightfall, St. James Street and Piccadilly were rivers of broken glass in which bits of jewelry, boots, clothing, and bottles bobbed.[82]

The riot sent a shudder of fear through London's wealthy West End. Though not a single life was lost in the riot and only a dozen rioters were arrested, most store owners complied with a police warning to keep their shops shuttered on Tuesday. A *New York Times* reporter derided the police's lack of preparedness—by Wednesday they were in a position to stop further riots should they occur, "what the police in Boston or New York would have promptly done—Monday afternoon"—sympathetically noting that this was the worst rioting London had seen since the infamous anti-Catholic riots of 1780.[83] Londoners agreed that there had not been looting on such a scale since Victoria took the throne nearly fifty years before, just after the first Reform Act passed.[84] The queen pronounced the riot "monstrous."[85]

The queen's assertion that the riot constituted "a momentary triumph for socialism" was almost certainly untrue.[86] But the episode did stimulate a good deal of activism and calls for action. Worried and conscience-stricken Londoners poured £79,000 into the Lord Mayor's relief fund for the unemployed and demanded that the money be dispensed. Beatrice's cousin Maggie Harkness began to plot a novel that she planned to call *Out of Work*.[87] Joseph Chamberlain, now a member of Prime Minister William Gladstone's new cabinet, set off a heated controversy by floating a public works scheme for the East End. Beatrice, exiled to the Potters' country estate and responsible not only for her father's care but also for a troubled younger sister and her father's equally troubled business affairs, was jolted out of her depression long enough to fire off a letter to the editor of the Liberal *Pall Mall Gazette* challenging the prevailing view of the causes and likely remedies for the crisis.

Beatrice braced herself for polite rejection. A letter from the journal's editor arrived by return post—too soon, she was sure, to contain any other message. But when she tore it open, she found a request for permission to publish "A Lady's View of the Unemployed" as an article under her byline. Beatrice shouted for joy. Her first real "bid for publicity" was a success; her thoughts and words had been judged worth listening to.[88] She had to believe that it was "a turning point in my life."[89]

Ten days after the riot, Beatrice had the pleasure of reading her own words in print for the first time: "I am a rent collector on a large block of working-class dwellings situated near the London Docks, designed and adapted to house the lowest class of working poor." She had tried to make just two points. Her first was that, contrary to what most philanthropists and politicians supposed, unemployment in the East End, "the great centre of odd jobs and indiscriminate charity," was the result not of "the national depression of trade" but of a dysfunctional and lopsided labor market. As traditional London trades such as shipbuilding and manufacturing had

moved away, record numbers of unskilled farm laborers and foreign immigrants had been attracted by false or exaggerated reports of sky-high
wages and jobs going begging. Her second point followed from the first:
advertising public works jobs would inevitably attract more unskilled
newcomers to the already overcrowded labor market, swelling the ranks of
the jobless and depressing the wages of those who had work.[90]

One week after her piece appeared, she was reading another letter that
made her heart pound and her hands shake. Chamberlain complimented
her article and wanted to solicit her advice. As president of the Local Government Board, he was now responsible for poor relief. Would she meet to
advise him on how to modify his plan to eliminate its pitfalls?[91] Her pride
still hurt and fearing further humiliation, Beatrice refused to meet Chamberlain and instead sent him a critique of his plan. Chamberlain's response
was a repetition of his "ransom" argument. As he put it, "the rich must
pay to keep the poor alive."[92] He had taken from his experiences as the
employer of thousands of workers the belief that government inaction in
the face of widespread distress was no longer an option. The rules of governing were changing, irrespective of which party was in power. As wealth
grew in tandem with the political power of the impoverished majority, a
moral and political imperative to act where none had existed previously
had emerged. Once the means to alleviate distress were available—and,
more important, once the electorate knew that such means existed—
doing nothing was no longer an option. Laissez-faire might have defined
the moral high ground in the poorer, agrarian England of Ricardo's and
Malthus's day, but any attempt to follow the precepts articulated in *The
Man Versus the State* in this day and age was immoral, not to mention politically suicidal. He wrote: "My Department knows all about Paupers . . .
I am convinced however that the suffering of the industrious non-pauper
class is very great . . . What is to be done for them?"[93]

Beatrice was unmoved. "I fail to grasp the principle that something
must be done," she persisted. Instead of proposing modifications, she
advised him to do nothing. "I have no proposal to make except sternness
from the state, and love and self-devotion from individuals," she wrote.
She could not resist adding, half mockingly, half flirtatiously, that

It is a ludicrous idea that an ordinary woman should be called upon to review the suggestion of her Majesty's ablest Minister, . . . especially when I know he has a slight opinion of even a superior woman's intelligence . . . and a dislike to any independence of thought.[94]

Chamberlain defended himself against her charges of misogyny and acknowledged that some of her objections were sound. Still, he did not disguise how repellent he found her underlying attitude:

On the main question your letter is discouraging; but I fear it is true. I shall go on, however, as if it were not true, for if we once admit the impossibility of remedying the evils of society, we shall all sink below the level of the brutes. Such a creed is the justification of absolute, unadulterated selfishness.[95]

Chamberlain did as he promised, ignoring Beatrice's advice and embarking on one of the "gigantic experiments" of which Spencer so disapproved. The public works program that Chamberlain pushed through was relatively modest in scale and lasted only a few months, but some historians judge it to have been a major innovation.[96] For the first time, government was treating unemployment as a social calamity rather than an individual failure and taking responsibility for aiding the victims.

When Chamberlain indicated that he was tired of their epistolary bickering, Beatrice impulsively fired off an angry confession that she loved him—to her instant and bitter regret. "I have been humbled as far down as a woman can be humbled," she told herself.[97] A doctor's suggestion that she take her father to London during the season saved her life. Instead of slipping back into her old depression and reaching for the laudanum bottle, she moved her household to York House in Kensington. Toward the end of April 1886, Beatrice joined her cousin Charlie Booth, a wealthy philanthropist, in the most ambitious social research project ever carried out in Britain.

Beatrice's cousin was in his forties, a tall and gawky figure with "the complexion of a consumptive girl" and a deceptively mild manner.[98] People who didn't know Charles Booth took him for a musician, professor, or priest—almost anything except what he was, the chief executive of a large transatlantic shipping company. By day he busied himself with share prices, new South American ports, and freight schedules. By night he turned to his real passions, philanthropy and social science. He and his wife, Mary, a niece of the historian Thomas Babington Macaulay, were an unpretentious, active, and intellectually curious couple. Political Liberals like the Potters and Heyworths, they were part of the "British Museum" crowd of journalists, union leaders, political economists, and assorted activists. Though Beatrice sometimes wrinkled her aquiline nose at the Booths' casual housekeeping and odd guests, she spent as much time at their haphazard mansion as she could.

Like other civic-minded businessmen, Booth had long been active in his local statistical society and shared the Victorian conviction that good data were a prerequisite for effective social action. When Chamberlain was mayor of Birmingham, he had done a survey at his behest and they had become friends. His finding that more than a quarter of Birmingham's school-age children were neither at home nor in school had led to a spate of legislation. In the early 1880s, when poverty amid plenty was again becoming the rallying cry for critics of contemporary society, he was struck by the widespread "sense of helplessness" that well-intentioned people felt in the face of an apparently intractable problem and a bewildering array of conflicting diagnoses and prescriptions. The trouble, he thought, was that political economists had theories and activists had anecdotes, but neither could supply an unbiased or complete description of the problem. It was as if he had been asked to reorganize South American shipping routes without the benefit of maps.

The previous spring, Booth had been outraged by an assertion by some Socialists that more than one-quarter of London's population was destitute. Suspecting but unable to prove that the figure was grossly exaggerated, he had been goaded into taking action. He determined that he would survey every house and workshop, every street and every type of employment, and learn the income, occupation, and circumstances of every one of London's 4.5 million citizens. Using his own money, he would create a map of poverty in London.

Unlike Henry Mayhew, whom Beatrice admired, Booth had the vision, managerial experience, and technical sophistication to carry out this extraordinary plan. His first step, after consulting friends such as Alfred Marshall, who was teaching at Oxford at the time, and Samuel Barnett of the settlement house Toynbee Hall, was to recruit a research team. Beatrice accepted his invitation to attend the first meeting of the Board of Statistical Research at the London branch of his firm. She was, of course, the only woman. Booth explained that he aimed to get a "fair picture of the whole of London society" and presented them with an "elaborate and detailed plan" that involved the use, among other things, of truant officers as interviewers and census returns and charity records as cross-checks.[99] He wanted to start with the East End, which contained 1 million out of London's 4 million inhabitants:

> My only justification for taking up the subject in the way I have done is that this piece of London is supposed to contain the most destitute population in England, and to be, as it were, the focus of the problem of poverty in the midst of wealth, which is troubling the minds and hearts of so many people.[100]

Beatrice was deeply impressed that Booth had launched the ambitious undertaking singlehandedly. She could imagine herself taking on a similarly pioneering role in the future. This was, she realized, "just the sort of work I should like to undertake . . . if I were free."[101] She decided to apprentice herself to her cousin, so to speak, devoting as much time and absorbing as much knowledge as caring for her family would allow. Her role was not going to be collecting statistics. Instead she was to go into workshops and homes, make her own observations, and interview workers—starting with London's legendary dockworkers.

When the Potters returned to their country estate, Beatrice took advantage of her enforced isolation to fill a gap in her education. Augmenting statistics with personal observation and interviews seemed essential to her, but she grasped that good observation was impossible without some theory to separate the wheat from the chaff. Mayhew had failed to produce lasting insights because he had gathered facts indiscriminately. The need for some sort of framework made her eager to learn some economics and especially to learn how economic ideas had evolved, since "each fresh development corresponded with some unconscious observation of the leading features of the contemporary industrial life." [102]

After a day or two of fitful reading, Beatrice complained that political economy was "most hateful drudgery." [103] A mere two weeks later, however, she was satisfied that she had "broken the back of economical science." [104] She had finished—or at least skimmed—Mill's *A System of Logic* and Fawcett's *Manual of Political Economy* and was convinced that she had "gotten the gist" of what Smith, Ricardo, and Marshall had to say. By the first week of August, she was putting the finishing touches on a critique of English political economy. Except for Marx, whose work she read in the fall, the leading political economists were guilty of treating assumptions as if they were facts, she argued, chiding them for paying too little attention to collections of facts about actual behavior. She sent her indictment to Cousin Charlie, hoping that he would help her get it published. To her chagrin, Booth wrote back suggesting that she put the piece away and return to it in a year or two.

A year later, after she had completed her study of dockworkers, Booth took Beatrice to an exhibition of pre-Raphaelite artists in Manchester. Beatrice was so moved by the paintings that she resolved to turn her next study— of sweatshops in the tailoring trades—into a "picture" too. It occurred to her that if she wanted to "dramatize" her account, she would have to go underground. "I could not get at the picture without living among the actual workers. This I think I could do." [105]

Preparing for her debut role as a working girl took months. She spent the summer at Standish, immersed in "all the volumes, Blue Books, pamphlets and periodicals bearing on the subject of sweating that I could buy or borrow." [106] In the fall, she lived in a small East End hotel for six weeks while she spent eight to twelve hours a day at a cooperative tailoring workshop, learning how to sew. At night, when she wasn't too exhausted to fall into her bed, she went out to fashionable West End dinner parties.

By April 1888, she was ready to begin her underground investigation. She moved to a shabby East End rooming house. The next morning she threw on a set of shabby old clothes and went off on foot "to begin life as a working woman." In a few hours, she got her first taste of job hunting.

It gave her, she confessed, "a queer feeling." As she wrote in her diary, "No bills up, except for 'good tailoress' and at these places I daren't apply, feeling myself rather an imposter. I wandered on, until my heart sank with me, my legs and back began to ache, and I felt all the feeling of 'out o' work.' At last I summoned up courage."[107]

"It don't look as if you have been 'customed to much work," she heard again and again. Still, twenty-four hours later, in spite of her fear that everyone saw through her disguise and her awkward attempts to drop her *h*'s, Beatrice was sitting at a large table making a clumsy job of sewing a pair of trousers. Her fingers felt like sausages, and she had to rely on the kindness of a fellow worker, who, though she was paid by the piece, took the time to teach Beatrice the ropes, and of the "sweater" who sent out a girl to buy the trimmings that workers were expected to supply themselves.

The woman whose motto was "A woman, in all the relations of life, should be sought," gleefully transcribed the lyrics of a work girls' song:

> *If a girl likes a man, why should she not propose?*
> *Why should the little girls always be led by the nose?*[108]

As soon as the gas was lit, the heat was terrific. Beatrice's fingers were sore and her back ached. "Eight o'clock by the Brewery clock," cried out a shrill voice.

For this she received a shilling, the first she had ever earned. "A shilling a day is about the price of unskilled women's labor," she recorded in her diary when she got back to the rooming house.

She was back at 198 Mile End Road at eight thirty the next morning. She sewed buttonholes on trousers for a couple of days before "leaving this workshop and its inhabitants to work on its way day after day and to become to me only a memory." [109]

News of Beatrice's exploit spread quickly. In May, a House of Lords committee that was conducting an investigation of sweatshops invited her to testify. The *Pall Mall Gazette,* which covered the hearing, described her in glamorous terms as "tall, supple, dark with bright eyes" and her manner in the witness chair as "quite cool." [110] In the hearing, Beatrice slipped into her childhood habit of fibbing and claimed that she had spent three weeks instead of three days in the sweatshop. Fear of exposure kept her in an agony of suspense for weeks afterward. But when "Pages of a Workgirl's Diary" was published in the liberal journal the *Nineteenth Century,* in mid-October, its success was delicious. "It was the originality of the deed that has taken the public, more than the expression of it." [111] All the same, Beatrice admitted, an invitation to read her paper at Oxford made her ridiculously happy. ("If I have something to say I now know I can say it and say it well." [112]) Just before New Year's, despite a bad cold that kept her in bed, Beatrice was luxuriating in mentions in the daily papers and "even a bogus interview . . . telegraphed to America and Australia." [113]

Beatrice now felt emboldened to embark on a project that was hers alone. Ever since her week as "Miss Jones" in Bacup among the hand-loom weavers, she had been drawn to the idea of writing a history of the cooperative movement. Even the shock of reading in the *Pall Mall Gazette* that Joseph Chamberlain had been secretly engaged to a twenty-five-year-old American "aristocrat"—"a gasp—as if one had been stabbed—and then it was over" [114]—did not prevent Beatrice from plunging into Blue Books once more. Her cousin Charlie tried to convince her to write a treatise about women's work instead. So did Alfred Marshall, whom she met for the first time at Oxford and who invited her to lunch with him and Mary. He greatly admired her "Diary," he said. When she seized the opportunity to ask him what he thought of her new project, he told her dramatically that "if you devote yourself to the study of your own sex as an industrial factor, your name will be a household word two hundred years hence; if you write a history of Co-operation, it will be superseded or ignored in a few years." [115]

Beatrice, who preferred spending her time with men rather than with other women, and who suspected that Marshall thought her unqualified to write about one of his favorite subjects, had no intention of taking such advice. The matter was clinched when she impulsively joined with other socially prominent women in signing a petition opposing female suffrage. "I was at that time known to be an anti-feminist," she later explained. [116]

In fact, Beatrice was changing her mind about a great many things. Notwithstanding her spirited defense to Chamberlain of her laissez-faire philosophy, she was beginning to have doubts about her parents' and Spencer's libertarian creed. She and the old philosopher still saw each other often, but their disagreements were now so violent that they talked less and less about politics. In any case, she was spending more and more of her time with her cousin Charlie.

When Booth published the first volume of *Labour and Life of the People* in April 1889, the *Times* said it "draws the curtain behind which East London has been hidden from view," and singled out Beatrice's chapter on the London dockworkers for praise. [117] In June of that year, Beatrice attended a cooperative congress, where she became convinced that "the democracy of Consumers must be complemented by democracies of workers" if workers could ever hope to enforce hard-won agreements on pay and working hours. [118] The dramatic and wholly unexpected victory in August 1889 of striking London dockworkers, universally believed to be too egotistical and desperate to band together, impressed her greatly. "London is in ferment: Strikes are the order of the day, the new trade unionism with its magnificent conquest of the docks is striding along," Beatrice wrote in her diary.

> The socialists, led by a small set of able young men (Fabian Society) are
> manipulating London Radicals, ready at the first check-mate of trade
> unionism to voice a growing desire for state action; and I, from the

peculiarity of my social position, should be in the midst of all parties, sympathetic with all, allied with none." [119]

Instead of witnessing these stirring sights firsthand, Beatrice was far away in a hotel in the country, tethered to her semicomatose father, "exiled from the world of thought and action of other men and women." She worked on her book, but without any conviction that she could ever complete it. She was "sick to death of grappling with my subject. Was I made for brain work? Is any woman made for a purely intellectual life? . . . The background to my life is inexpressibly depressing—Father lying like a log in his bed, a child, an animal, with less capacity for thought and feeling than my old pet, Don." [120]

As her frustration grew at being unable to sustain a career while nursing her father, Beatrice was more and more inclined to identify the plight of women with the oppression of workers. She thought about the houses of "all those respectable and highly successful men" and her sisters, to whom she remained close, had married:

> Then . . . I struggle through an East End crowd of the wrecks, the waifs
> and strays, or I enter a debating society of working men and listen to the
> ever increasing cry of active brains doomed to the treadmill of manual
> labor—*for a career in which ability tells*—the bitter cry of the nineteenth
> century working man and the nineteenth century woman alike. [121]

The previous fall, when her father had told her, "I should like to see my little Bee married to a good strong fellow," Beatrice wrote in her journal, "I cannot, and will never, make the stupendous sacrifice of marriage." [122]

Beatrice became aware of Sidney Webb months before she met him. She read a book of essays published by the Fabian Society, a Socialist group that intended to win power the same way a Roman general named Fabius had won the Carthaginian war; gradually and with guerrilla tactics rather than head-on battles. She told a friend that "by far the most significant and interesting essay is by Sidney Webb." [123] Sidney returned the compliment in a review of the first volume of the Booth survey: "The only contributor with any literary talent is Miss Beatrice Potter."[124]

Their first encounter took place in Maggie Harkness's rooms in Bloomsbury. Beatrice had asked her cousin if she knew of any experts on cooperatives, and Maggie immediately thought of a Fabian who seemed to know everything. For Sidney, it was love at first sight, though he left their first meeting more despondent than euphoric. "She is too beautiful, too rich, too clever," he said to a friend.[125] Later he comforted himself with the thought that they belonged to the same social class—until Beatrice corrected him. True, blue-collar men amused her. She enjoyed talking and smoking with union activists and cooperators in their cramped flats. But the self-importance of working men who, having "risen . . . within their own class," show up at London dinners and "introduce themselves as such without the least uneasiness for their reception" provoked her inner snob.[126] Beatrice thought Sidney looked like a cross between a London cardsharp and a German professor and mocked his "bourgeois black coat shiny with wear" and his dropped *h*'s. Unaccountably, she found that something about this "remarkable little man with a huge head on a tiny body" appealed to her.[127]

As his "huge head" indicated, Sidney was indeed a great brain. Like Alfred Marshall, he was very much a son of London's lower-middle class and had risen on the tide that was lifting white-collar workers. Born three years after Beatrice, he grew up over his parents' hairdressing shop near Leicester Square. His father, who moonlighted as a freelance bookkeeper in addition to cutting hair, was a radical democrat who had supported John Stuart Mill's parliamentary campaign. Sidney's mother, who made all the important decisions in the family, was determined that Sidney and his brother would grow up to be professionals. With a prodigious memory, a head for numbers, and a talent for test taking, Sidney excelled in school, got hired by a stockbroker at sixteen, and was offered a partnership in the firm at twenty-one. Instead of accepting, he took a civil service examination and won an appointment in the Colonial Office. By then he had been bitten by the political bug and realized that he was more interested in power than in money. He continued to collect scholarships and degrees including one in law from the University of London, according to the Webbs' official biographer, Royden Harrison. By the time of the Trafalgar Square Riot and the subsequent Tory electoral victory, Sidney had found his true vocation as the brains of the Fabian Society.

The Fabians were odd ducks. Sidney embraced "collective ownership where ever practicable; collective regulation everywhere else; collective provision according to need for all the impotent and sufferers; and collective taxation in proportion to wealth, especially surplus wealth." But Fabian Socialism was associated mostly with local government and small-scale projects such as dairy cooperatives and government pawnshops. The Fabians' strategy differed from that of most other Socialist groups as well. Eschewing both electoral politics and revolution, they sought to introduce Socialism gradually by "impregnating all the existence of forces of society with Collectivist ideals and Collectivist principles." [128]

When Sidney was elected to the Fabian steering committee in 1887, the society had sixty-seven members, an annual income of £32, and a reputation for being a good place for pretty women to meet brilliant men and vice versa. The English historian G. M. Trevelyan described the Fabians as "intelligence officers without an army." They did not aspire to become a political party in Parliament. Instead, they hoped to influence policies, "the direction of the great hosts moving under other banners." [129] Sidney, who had concluded that "nothing in England is done without the consent of a small intellectual yet practical class in London not 2,000 in number" and that electoral politics was a rich man's game, called the Fabian strategy of infiltrating the establishment "permeating." [130]

Sidney's best friend and partner in crime was George Bernard Shaw, a witty Irish sprite of a man who dashed off theater reviews and acted as the Fabians' chief publicist. By the mid-1890s, the former Dublin rent collector and City of London stockbroker was convinced that social problems had economic origins. He proceeded to devote the second half of the 1890s to "mastering" economics. He and Sidney were both trying to work out what they believed and where to direct their energies. They attended regular meetings of a group organized by several professional economists at City of London College. Their studies led them to reject both utopian

Socialism and Marxist Communism. They called their goal Socialism, but it was Socialism with property, Parliament, and capitalists and without Marx or class warfare. They wished to tame and control the "Frankenstein" of free enterprise rather than to murder it, and to tax the rich rather than to annihilate them.[131]

Within a few weeks of meeting Sidney for the first time, Beatrice was beginning to think that "a socialist community in which there will be individual freedom and public property" might be viable—and attractive. "At last I am a socialist!" she declared.[132] Beatrice had caught the spirit of the times that prompted William Harcourt, a Liberal MP, to exclaim during the 1888 budget debate, "We are all socialists now."[133] As for Sidney, she was beginning to think of him as "one of the small body of men with whom I may sooner or later throw in my lot for good and all."[134]

At first, Beatrice had taken Sidney's obvious infatuation for granted and had been happy to let her intellectual dependence on him grow. When he confessed that he adored her and wanted to marry her, she had responded with a lecture against mixing love with work. She had insisted on being his collaborator, not his wife, and had banned any further allusion to "lower feelings."[135]

In 1891, Beatrice was again living in London for the season, nervously waiting for her book on cooperatives to appear in print and worrying about a series of lectures she had agreed to deliver. Sidney announced that he was quitting the civil service. He had no life other than work and felt "like the London cabhorse who could not be taken out of his shafts lest he fall down."[136] He once again brought up the forbidden topic, promising that if she relented he would let her live the outdoorsy, abstemious, hardworking, intensely social life she wanted. He suggested they write a book on trade unions together. After a year of telling Sidney "I do not love you," Beatrice finally said "yes."[137]

When Sidney sent Beatrice a full-length photograph of himself, she begged him to "let me have your head only—it is your head only that I am marrying . . . It is too hideous for anything."[138] She dreaded telling her family and friends. "The world will wonder," Beatrice wrote in her diary.

On the face of it, it seems an extraordinary end to the once brilliant Beatrice Potter ... to marry an ugly little man with no social position and less means, whose only recommendation so some may say is a certain pushing ability. And I am not "in love," not as I was. But I see something else in him ... a fine intellect and a warmheartedness, a power of self-subordination and self-devotion for the common good.[139]

Beatrice insisted that the engagement was to be a secret as long as her father was alive. Only her sisters and a few intimate friends were told. The Booths reacted coolly, and Herbert Spencer promptly dropped her as literary executor, a position that had once been a source of great pride to Beatrice.

Richard Potter died a few days before Beatrice's thirty-fourth birthday on New Year's Day, 1892. He bequeathed his favorite daughter an annual income of £1,506 a year and "incomparable luxury of freedom from all care."[140] After the funeral, Beatrice spent a week at her prospective mother-in-law's "ugly and small surroundings" in Park Village near Regents Park. On July 23, 1892, Beatrice and Sidney were married in a registry office in London. Beatrice recorded the event in her diary: "Exit Beatrice Potter. Enter Beatrice Webb, or rather (Mrs.) Sidney Webb for I lose alas! both names."[141]

When George Bernard Shaw paid his first extended visit to the newlyweds more than a year later, in the late summer of 1893, Beatrice sized him up as vain, flighty, and a born philanderer, but "a brilliant talker" who "liked to flirt and was therefore a delightful companion." While Sidney was the "organizer" of the Fabian junta, she put Shaw down as its "sparkle and flavour."[142]

Shaw's first play, *Widowers' Houses*, had been put on at the Royalty Theatre the previous December, and now he was at work on a new play that operated on the same formula: taking one of Victorian society's "unspeakable subjects," in this case a reviled profession, and turning it into a metaphor for the way the society really worked.[143]

All the past year the press had been full of stories about the Continent's legal brothels—high-end men's clubs where business was conducted—in which English girls were lured into sexual slavery. As usual, Shaw was recasting a social problem as an economic problem, and he wrote to another friend that "in all my plays my economic studies have played as important a part as knowledge of anatomy does in the works of Michael Angelo." [144] His character Mrs. Warren, who runs a high-end brothel in Vienna, is a practical businesswoman who understands that prostitution isn't about sex but about money. Just as he had wanted the audience to see that the slumlord of *Widowers' Houses* was not a villain but a symptom of a social system in which everyone was implicated, he now wished them to understand that in a society that drives women into prostitution, there were no innocents. "Nothing would please our sanctimonious British public more than to throw the whole guilt of Mrs. Warren's profession on Mrs. Warren herself," wrote Shaw in a preface, "Now the whole aim of my play is to throw that guilt on the British public itself." [145]

It was Beatrice who suggested that Shaw "should put on the stage a real modern lady of the governing class" rather than a stereotypical sentimental courtesan. [146] The result was Vivie Warren, the play's heroine and Mrs. Warren's Cambridge-educated daughter. Like Beatrice, Vivie is "attractive ... sensible ... self-possessed." Like Beatrice, Vivie escapes her class and sexual destiny. In the Guy de Maupassant story "Yvette," which supplied Shaw with his plot, birth is destiny. "There's no alternative," says Madame Obardi, the prostitute mother to Yvette, heroine of the story, but in the world that Vivie Warren inhabits—late Victorian England—there *is* an alternative. The discovery of Mrs. Warren's real business and the true source of the income that had paid for her daughter's Cambridge education shatter Vivie's innocence. But instead of killing herself or resigning herself to following in her mother's footsteps, Vivie takes up ... accounting. "My work is not your work, and my way is not your way," she tells her mother. As with Beatrice, the choice not to repeat history was hers. In the final scene of *Mrs. Warren's Profession,* Vivie is alone onstage, at her writing desk, buried luxuriously in her "actuarial calculations."

Meanwhile, the real-life Vivie was living with her husband in a ten-room house a stone's throw from Parliament. She was joined in the library nearly every morning by Sidney and Shaw. The three of them drank coffee, smoked cigarettes, and gossiped while they edited the first three chapters of her and Sidney's book on trade unions.

Herbert George Wells, the wildly popular science-fiction writer, turned the Fabian trio into a quartet for a while before falling out with the Webbs. Afterward he satirized them in his 1910 novel *The New Machiavelli,* as Altiora and Oscar Bailey, a London power couple who steadfastly acquire and publish knowledge about public affairs in order to gain influence as the "centre of reference for all sorts of legislative proposals and political expedients." Having grown up among the ruling class like Beatrice, Altiora "discovered very early that the last thing influential people will do is work." Indolent but brilliant, she marries Oscar for his big forehead and industrious work habits, and under her steerage they become "the most formidable and distinguished couple conceivable." "Two people . . . who've planned to be a power—in an original way. And by Jove! They've done it!" says the narrator's companion.[147]

The term *think tank,* which connotes the growing role of the expert in public policy making, wasn't coined until World War II. Even then, according to the historian James A. Smith, *think tank* referred to a "secure room in which plans and strategies could be discussed."[148] Only in the 1950s and 1960s, after Rand and Brookings became familiar names, was *think tank* used to evoke private entities employing researchers, presumably independent and objective, that dispensed free, nonpartisan advice to civil servants and politicians. Yet a think tank is exactly what Beatrice and Sidney were—perhaps the very first and certainly one of the most effective—from the moment they married. "Of this they were unself-consciously proud," mocked Wells. "The inside of the Baileys' wedding rings were engraved 'P.B.P., Pro Bono Publico.'"

The Webbs shrewdly realized that experts would become more indispensable the more ambitious democratically elected governments became. They shared the vision of a new mandarin class: "From the mere necessities of convenience elected bodies *must* avail themselves more and more of the services of expert officials . . . We want to suggest that these expert officials must necessarily develop into a new class and a very powerful class . . . We consider ourselves as amateur unpaid precursors of such a class." [149] This insight led them to found the London School of Economics, intended as a training ground for a new class of social engineers, and the *New Statesman* weekly newspaper.

Their "almost pretentiously matter-of-fact and unassuming" house at 41 Grosvenor Road, chosen by Beatrice, advertised their priorities. To stay fit, their daily regimen was Spartan. Middle-class comfort was sacrificed for the sake of books, articles, interviews, and testimony. In an era of coal scuttles and cold running water, the Webbs generally employed three research assistants but only two servants. "All efficient public careers," says Altiora in Wells's novel, "consist in the proper direction of secretaries." [150] Beatrice set for herself the task of converting England from laissez-faire to a society planned from the top down. To this end, they plotted ambitious research projects and organized their lives almost entirely to meet deadlines. The Webbs' friends debated "as to which of the two is before or after the other," but according to Wells, "[S]he ran him." [151] She was the CEO of the Webb enterprise; part visionary, part executive, and part strategist. Wells was sure that their joint career as idea brokers was "almost entirely her invention." In his view, Beatrice was "aggressive, imaginative, and had a great capacity for ideas" while Sidney "was almost destitute of initiative and could do nothing with ideas except remember and discuss them." [152]

Standing with her back to the fire, Beatrice glowed with "a gypsy splendor of black and red and silver all her own." Even while caricaturing her in his novel, Wells was forced to admit that Beatrice was beautiful, elegant, and "altogether exceptional." The other women he had met at Grosvenor House were either "severely rational or radiantly magnificent." [153] Beatrice was the only one who was both. Even as she talked of budgets, laws, and political machinations, she signaled her femininity by wearing outrageously expensive, flirty shoes.

A daddy's girl, Beatrice had always adored powerful men, flirting, and political gossip. The Fabians' strategy of permeation gave her an excuse to indulge all three. "I set myself to amuse and interest him, but seized every opportunity to insinuate sound doctrine and information" is a typical account of dining with a prime minister. Past, present, and future prime ministers were among her regular celebrity guests. Not in the slightest partisan, she was as happy to entertain a Tory as a Liberal. "But all of these have certain usefulness," she observed pragmatically.[154]

The think tank became a political salon at night. Once a week, the Webbs had a dinner for a dozen or so people. Once a month, they had a party for sixty or eighty. Guests did not come for the food. The Webbs practiced strict household economy to afford more research assistants, and Beatrice took more satisfaction in disciplining than in indulging her appetite.[155] Like Altoria, Beatrice fed her guests "with a shameless austerity that kept the conversation brilliant."[156] The price of attendance, said R. H. Tawney, an economic historian and frequent guest, was "participation in one of the famous exercises in asceticism described by Mrs. Webb as dinners."[157] Yet everyone angled for invitations, and 41 Grosvenor Road was a center for quite an astonishing amount of political and social activity. One "brilliant little luncheon" that Beatrice regarded as "typical of the 'Webb' set . . . in its mixture of opinions, classes, interests" included the Norwegian ambassador to London, a Tory MP, a Liberal MP, George Bernard Shaw, Bertrand Russell, the philosopher and future Nobel laureate, and a baroness who entertained every major politician and writer of the time.[158] Wells's novel identified Beatrice's singular skill as a hostess and its importance to the Webbs' career. "She got together all sorts of interesting people in or about the public service. She mixed the obscurely efficient with the ill-instructed famous and the rudderless rich, and got together in one room more of the factors in our strange jumble of a public life than had ever met easily before."[159]

A first-time visitor to the Baileys' tells the friend who has brought him, "'It's the oddest gathering.'"

"'Every one comes here,'" says the regular. "'Mostly we hate them like poison—jealousy—and little irritations—Altiora can be a horror at times—but we *have* to come.'"

"'Things are being done?'" asks the first man.

"'Oh!—no doubt of it. It's one of the parts of the British machinery—that doesn't show.'" [160]

Winston Churchill was one of those who *had* to come during the 1903 London season. He had been seated next to Beatrice at a Liberal dinner the previous year. Then the scion of an old aristocratic family, the Spencers, and son of a famous former Tory politician, and now a Tory member of Parliament, Churchill was thought to be at odds with the Conservative government. But he irritated Beatrice by declaring his opposition not just to trade unions but to public elementary school education. Worse, he talked about himself without pause from drinks to dessert, addressing Beatrice only to ask whether she knew someone who could get him statistics. "I never do my own brainwork that anyone else can do for me," he said breezily. "Egotistical, bumptious, shallow-minded and reactionary," Beatrice scrawled angrily in her diary that night. There is no record of his reaction to her. [161]

By the time Churchill reappeared chez Webb, he had defected to the Liberal opposition. The mood of the electorate was changing. After the costly and futile war against the Boers in South Africa, British voters were disillusioned with imperialism abroad and anxious about poverty at home. The Tories, who had been the ruling party for nearly a decade—first under the Marquess of Salisbury, then Arthur Balfour—proposed a protectionist plank but succeeded only in alienating working-class voters who feared higher food prices and lost jobs in export industries; Joseph Chamberlain, who drafted the Tory tariff "reform" program, was making the final speeches of his political career to virtually empty halls. Alfred Marshall, who had come out of retirement to blast Chamberlain and the protectionists, wondered if the distress of becoming entangled in a public controversy had even been necessary. Churchill was quick to sense the Tories' growing irrelevance and thought that the Liberals were ready to move to the left with the rest of the nation. He took this to mean that they had to address the social question . . . somehow. Without trade union votes, he reasoned, the Liberals had no chance of remaining in power, presuming they could get voted in to begin with.

At dinner, Beatrice sat Churchill to her right. He managed to make almost as bad an impression as the first time. The woman who had just decided to forswear not only alcohol but coffee and tobacco (tea remaining her "one concession to self-indulgence") came away convinced that "he drinks too much, talks too much, and does no thinking worthy of the name." She discussed the idea of a guaranteed "national minimum" standard of living with Churchill. He merely trotted out what she called "infant school economics." Her verdict: "He is completely ignorant of all social questions . . . and does not know it . . . He is evidently unaware to the most elementary objections to unrestricted competition." [162]

Near the end of his magisterial history of nineteenth-century England, the French historian Élie Halévy mentions several pieces of legislation of "almost revolutionary importance . . . passed on Churchill's initiative." [163] Among these measures was the "first attempt to introduce a minimun wage into the Labor code of Great Britain, which formed part of the Webbs' formula for the 'National Minimum.'"

Although Churchill found Beatrice overbearing—"I refuse to be shut up in a soup kitchen with Mrs. Sidney Webb," he later said—he was in fact aware of his own ignorance and soon began "living with Blue Books and sleeping with encyclopedias." [164] While he and Beatrice saw little of each other, Churchill plowed through most of the Fabian syllabus, from Booth's *Life and Labours* and Seebohm Rowntree's *Poverty: A Study of Town Life* to Beatrice and Sidney's *History of Trade Unionism* and *Industrial Democracy.* H. G. Wells, whose subject matter was shifting from science fiction to social engineering, became his favorite novelist. "I could pass an examination in them," Churchill boasted.[165] A great fan of Shaw's, he attended the opening of *Major Barbara.* At one point, he and his personal secretary, Eddie Marsh, spent hours wandering through some of Manchester's worst slums, just as Alfred Marshall had a generation earlier. "Fancy living in one of these streets—never seeing anything beautiful—never eating anything savory—*never saying anything clever!*" Churchill said to Marsh afterward.[166]

Such was Churchill's shock, reports his biographer William Manchester, that before long the former archconservative had become "a thunderer on the left." His inspirations were numerous, and political calculation played a role, but the specific arguments and remedies were mostly borrowed from Beatrice. By early 1906, when the Liberals won a majority by a landslide, Churchill was preaching what he called the "cause of the left-out millions" and urging "drawing a line" below which "we will not allow persons to live and labor"—precisely the policy that Beatrice had been urging on him.[167]

That October, Churchill gave a remarkable speech in Glasgow that not only went far beyond what the leaders of the Liberal Party had in mind but, according to the Churchill biographer Peter de Mendelssohn, "contained the nucleus of many essential elements of the programme with which the Labour Party obtained its overwhelming mandate for the 'silent revolution' of 1945–50."[168] In one of his most brilliant rhetorical performances, Churchill argued that the "whole tendency of civilization was towards the multiplication of the collective functions of society," which he believed rightly belonged to the state rather than private enterprise:

> I should like to see the State embark on various novel and adventuresome experiments. . . . I am of opinion that the State should increasingly assume the position of the reserve employer of labour. I am very sorry we have not got the railways of this country in our hands . . . and we are all agreed . . . that the State must increasingly and earnestly concern itself with the care of the sick and the aged, and above all, of the children. I look forward to the universal establishment of minimum standards of life and labour, and their progressive elevation as the increasing energies of production may permit . . . I do not want to see impaired the vigor of competition, but we can do much to mitigate the consequences of failure. . . . We want to have free competition upwards; we decline to allow free competition downwards. We do not want to pull down the structure of science and civilizations; but to spread a net over the abyss.[169]

No one has a greater claim to the invention of the *idea* of a government safety net—indeed, the modern welfare state—than Beatrice Webb. Looking back shortly before her death in 1943, she noted with satisfaction, "We saw that to the Government alone could be entrusted the provision for future generations . . . In short, we were led to the recognition of a new form of state, and one which may be called the 'house-keeping state,' as distinguished from the 'police state.'" [170]

The germ of the idea grew out of her and Sidney's study of trade unions. In their 1897 book *Industrial Democracy,* they had proposed sweeping national health and safety standards. A "national minimum" would shelter the entire workforce except for farm laborers and domestic servants. The most radical component was a national minimum wage. Arguing that "in the absence of regulation, the competition between trades tends to the creation and persistence in certain occupations of conditions of employment injurious to the nation as a whole," they insisted that a government-imposed floor under pay and working conditions was not, as Marx and Mill had assumed, inherently incompatible with the unimpeded productivity growth on which gains in real wages and living standards depended. [171] Indeed, they claimed, the cost to business of the regulations would be more than offset by fewer industrial accidents and a better-nourished, more alert workforce. Still, they admitted that the huge expansion of government power over private enterprise went far beyond anything that the trade union leaders, who mainly wanted a free hand to fight for higher pay and better working conditions, had in mind.

But the more ambitious idea of "a new form of state" didn't seize Beatrice until nearly a decade later. At the end of 1905, in the last days of the Tory Balfour government, she was appointed to a royal commission to reform the Poor Laws. The commission continued for three years under the new Liberal government. From the beginning, Webb clashed with the other commissioners. Seizing on Alfred Marshall's suggestion that "the cause of poverty is poverty," she defined the problem in absolute rather than relative terms. Inequality, and therefore poverty in the sense of having less than others, is inevitable, she reasoned, but destitution, "the condition of being without one or other of the necessaries of life, in such a way that health and strength, and even vitality, is so impaired as to eventually imperil life itself," is not. [172] Eliminating destitution would prevent the poverty of one generation from passing automatically to the next.

From her East End days, she could speak authoritatively about families in which "now in one and now in another of its members, sores, indigestion, headaches, rheumatism, bronchitis and bodily pains alternate almost ceaseless, to be periodically broken into by serious disease, and cut short by premature death"; or families where the father is out of work, "meaning as it does lack of food, clothing, firing, and decent housing conditions"; or about those who could not work: widows with little children, the aged, or the lunatic.[173]

Webb dismissed the notion that destitution could always be traced to a moral defect. Instead she listed five causes that corresponded to the main groups of destitute individuals and families: the sick, widows with young children, the aged, and people suffering from a variety of mental afflictions, from low intelligence to lunacy. The most troubling group were the able-bodied destitute. Their destitution, Beatrice argued, was the result of unemployment and chronic underemployment.

She made it clear that the urgent need to eliminate destitution didn't arise "from any sense that things are getting worse, but because our standards are, in all matters of social organization, becoming steadily higher," by which she meant both that the working classes now had the vote and that Britain's main international competitor, Germany, had adopted a variety of social welfare measures.[174]

The problem of Britain's existing policy was that it offered relief only to those desperate enough to seek it and did nothing to prevent destitution and dependency in the first place. As Beatrice put it, "all these activities of the Poor Law Authority in relieving the destitution of the sweated worker, did nothing to prevent sweating" or "saving men and women from being thrown out of work or in warding off the oncoming of illness . . . [stopping] the unnecessary killing and maiming of the workers by industrial accidents, or the wanton destruction of their health by insanitary housing and preventable industrial diseases."[175]

She wanted the government as much as possible to get out of the business of dispensing welfare and into the business of eliminating poverty's causes. "The very essence of the Policy of Prevention that what has to be supplied in every case is not relief but always treatment and the treatment appropriate to the need." [176] She never questioned whether the government or its experts knew how to treat the "disease of modern life" or worried about its cost. Inevitably, her ambitious vision of a "housekeeping state" that prevented rather than merely relieved poverty clashed with the more limited aims of the other commission members. As she had planned all along, she refused to sign the commission's report. Instead, she and Sidney spent the first nine months of 1908 pouring her vision into a document called *The Minority Report*, which she convinced three other commissioners to sign. Their "great collectivist document," as she called it,[177] envisioned a cradle-to-grave system designed to "secure a national minimum of civilized life . . . open to all alike, of both sexes and all classes, by which we meant sufficient nourishment and training when young, a living wage when able-bodied, treatment when sick, and a modest but secure livelihood when disabled or aged." [178]

Webb conceded that the idea would be regarded as utopian by other reformers and amounted to a repudiation of traditional, limited government. In contrast to the Socialist state, she believed, the household state was perfectly compatible with free markets and democracy. Indeed, she presented the welfare state as merely the next stage in the natural evolution of the liberal state. Yet the notion that the basic welfare of citizens was the responsibility of the state and that the government was obliged to guarantee a minimum standard of living to every citizen who could not provide it for himself was not only a departure from Spencer's ideal of a minimal state. Beatrice's idea broke with the whole tradition of Gladstonian liberalism that promised equality of opportunity but left results up to the individual and the market and went far beyond anything being discussed at the time by anyone except the Socialist fringe.

"It may make as great a difference in sociology and political science as Darwin's *Origin of Species* did in philosophy and natural history," her friend George Bernard Shaw predicted in his review of *The Minority Report*. "It is big and revolutionary and sensible and practical at the same time, which is just what is wanted to inspire and attract the new generation." He went on: "His right to live and the right of the community to his maintenance in health and efficiency are seen to be quite independent of his commercial profit for any private employer." That is, the objectives went far beyond Marshall's notion of increasing productivity and pay. "He is a cell of the social organism, and must be kept in health if the organism is to be kept in health." [179]

Ideas such as the minimum wage or minimum standards for leisure, safety, and health for all workplaces, the safety net, employment offices, fighting cyclical unemployment by shifting the timing of large government projects—basically, the whole notion that not only are the conditions that produce chronic poverty, or the more acute condition Webb called destitution, preventable, but it is the government's job to prevent them, and in order to prevent them the government must acquire new capabilities—have multiple authors. But nobody expressed these ideas so clearly, so systematically, or so often directly to the "mendicants of practicable proposals." And no one else found a phrase that made revolutionary changes seem evolutionary, even inevitable.

Making radical change seem evolutionary was Beatrice's genius. Even she, however, was surprised at how quickly ideas she and Sidney had thought were utopian in the 1890s seemed practical, or at least politically relevant, a decade later. Looking back at *Industrial Democracy* years later, she remarked with some measure of satisfaction, "What, in fact has characterized the social history of the present century has been the unavowed and often perfunctory adoption, in administration as well as in legislation of the policy of the National Minimum, formulated in this book." [180]

The year 1908 was a pivotal one for the new Liberal government. With unemployment and trade union militancy on the rise, an overwhelming Liberal majority in Parliament, and the "social problem" at the top of the political agenda, there was a general "scramble for new constructive ideas," Beatrice reported in her diary. The Webb's stock was soaring. "We happen just now to have a good many [ideas] to give away, hence the eagerness for our company," Beatrice continued happily. "Every politician one meets wants to be 'coached.' It is really quite comic. It seems to be quite irrelevant whether they are Conservatives, Liberals or Labor Party men—all alike have become mendicants for practicable proposals."[181] This justified a splurge, she decided, and ordered a new evening gown.

"Winston has mastered the Webb scheme," Beatrice crowed in October 1908 and remarked that they had "renewed our acquaintance." Having risen to her challenge, Churchill could now be classified as "brilliantly able—more than a phrase monger" in Beatrice's diary.[182]

For the first two years of the Liberal government led by Herbert Henry Asquith, Churchill's reforms had amounted to little more than rhetoric. Despite their electoral landslide in 1906, the Liberals had managed to push through very little of their program beyond restoring certain protections to trade unions. The logjam was broken in April 1908 when the thirty-three-year-old Churchill succeeded Lloyd George as president of the Board of Trade, a cabinet-level appointment. Beatrice found the cabinet shuffle "exciting."[183] The position, which combined many of the duties of the U.S. Departments of Labor and Commerce, entailed a grab bag of responsibilities: patent registration, company regulation, merchant shipping, railways, labor arbitration, and advising the foreign office on trade matters. Ultimately, Lloyd George's biographer points out that the president's responsibilities boiled down to ensuring "the smooth, orderly working of capitalism."[184] But Churchill proceeded to use the post to introduce radical social reforms. Remarked one of his friends at the time: "He is full of the poor whom he has just discovered. He thinks he is called by Providence to do something for them. 'Why have I always been kept safe within a hair's breadth of death,' he asked, 'except to do something for them?'"[185]

For the next two years, Churchill and Lloyd George, now Chancellor of the Exchequer, formed a partnership that ended, once and for all, "the old Gladstonian tradition of concentrating on libertarian political issues and leaving 'the condition of the people' to look after itself."[186] The new president of the Board of Trade did not wait to be sworn in before sitting up an entire night and writing the prime minister a long letter outlining his personal policy wish list. After the briefest of rhetorical flourishes—

"Dimly across gulfs of ignorance I see the outline of a policy which I call the Minimum Standard"[187]—Churchill defined his minimum in terms of five elements and listed them as his legislative priorities: unemployment insurance, disability insurance, compulsory education to age seventeen, public works jobs in road building or state afforestation in lieu of poor relief, and nationalization of the railways.

The recession that followed the panic of 1907 gave Churchill's proposals immediacy. Unemployment among trade union members, which was 5 percent at the end of 1907, doubled within a year. Alfred Marshall had shown that rising unemployment was usually caused by falling business activity. Now Beatrice demonstrated that unemployment, in turn, was a major cause of poverty. There was, however, no consensus that the government should, or could, intervene. Churchill intended to challenge the conventional wisdom. Aware that what he proposed far exceeded anything Asquith, the prime minister, had in mind, he urged the Liberal government to follow Germany's example and introduce unemployment and health insurance: "I say—thrust a big slice of Bismarckianism over the whole underside of our industrial system, & await the consequences whatever they may be with a good conscience."[188] He "is definitely casting in his lot with the [cause of] constructive state action,"[189] Beatrice exulted, while concluding that "Lloyd George and Winston Churchill are the best of the [Liberal] party."[190] She appreciated Churchill's "capacity for quick appreciation and rapid execution of new ideas, whilst hardly comprehending the philosophy beneath them."[191]

Eventually the whole reform effort was swallowed up by the Liberals' struggle to wrest the veto from the House of Lords. What is remarkable is how much of it was passed, remarks William Manchester: "Before the ascendancy of Churchill and Lloyd George, all legislative attempts to provide relief for the unfortunate had failed."[192]

Webb lost the battle over social insurance, which was far less expensive than direct provision of services. But ultimately she won the war of the welfare state. She and Sidney had provided the rationale for the "assumption by the state of responsibility for an increasing number of services, administered by a growing class of experts, and supported by an expanded apparatus of the state."[193] *The Minority Report* was one of the first descriptions of the modern welfare state. Lord William Beveridge, the eponymous author of the 1942 Beveridge Plan who worked on *The Minority Report* as a researcher, later acknowledged that his design for the post–World War II British welfare state "stemmed from what all of us had imbibed from the Webbs."[194]

Chapter IV

Cross of Gold:
Fisher and the Money Illusion

These dear people, always starting their new experiments in such
deadly earnest; taking themselves so seriously; really believing
that they are getting better and better, wiser and wiser—certain
that they are getting richer and richer—every year, every month,
every day . . . Oh! It is fine, Mrs. Webb: it is fine.

—*H. Morse Stephens*[1]

"To America . . . when they might go to Russia, India or China. What
taste!" a Tory acquaintance sneered when Beatrice and Sidney announced
that they were off to New York in the spring of 1898.[2] As the putdown
implied, the Webbs were traveling not as tourists but social investigators.
Nonetheless, Beatrice went on a shopping spree, scooping up "silks and
satins, gloves, underclothing, furs and everything a sober-minded woman
of forty can want to inspire Americans and colonials with a true respect for
the refinements of collectivism."[3] If she was going to tour the world's social
laboratory, she intended to dazzle the locals.

The Americanization of the World did not become a best seller for another year or two, but the Webbs were surely familiar with the views of its
author, William Stead, editor of the *Pall Mall Gazette*. Stead was convinced
that Britain's economic future was tied to its former colony. The two
economies were more intertwined than in the eighteenth century, when
America was a British dominion, or in the 1860s, when the Union block-

ade of Southern ports during the American Civil War led to the terrible cotton famine in Lancashire. In the last quarter of the nineteenth century, imperial preferences notwithstanding, Britain imported more raw materials from the United States than from her own colonies.[4] The term "American invasion" was coined by British journalists more than half a century before the French revived it in the 1960s.[5]

In 1902, a London newspaper complained:

The average citizen wakes in the morning at the sound of an American alarum clock; rises from his New England sheets, and shaves with his New York soap, and Yankee safety razor. He pulls on a pair of Boston boots over his socks from West Carolina, fastens his Connecticut braces, slips his Waterbury watch into his pocket and sits down to breakfast . . . Rising from his breakfast table the citizen rushes out, catches an electric tram made in New York, to Shepherds Bush, where he gets into a Yankee elevator, which takes him on to the American-fitted railway to the city. At his office of course everything is American. He sits on a Nebraska swivel chair, before a Michigan roll-top desk, writes his letter on a Syracuse typewriter, signing them with a New York fountain pen, and drying them with a blotting sheet from New England. The letter copies are put away in files manufactured in Grand Rapids.[6]

William Gladstone, the Liberal prime minister, had long predicted that the United States would inevitably wrest commercial supremacy from Britain. "While we have been advancing with this portentous rapidity," he observed in 1878, "America is passing us by as if in a canter."[7] In 1870, the most basic yardstick of average living standards, national income (GDP per person), was 25 percent higher in Britain. But for the next thirty years, the most basic measure of an economy's productive power and the key determinant of the average level of wages, nation income GDP per worker, rose nearly twice as fast in the United States.[8] One reason was that British citizens invested more than half their annual savings in America each year, more than at home and many times as much as in neighboring European countries.[9] The earnings from these investments added to Britain's national income in any given year, while the investments themselves enabled American business to modernize. Another reason was that the United States was the chosen destination of more than half of all British and an even larger fraction of Irish emigrants, almost 8 million men, women, and children over three decades. Canada, by contrast, attracted fewer than 15 percent of British emigrants even though its culture felt more "English."[10] Average incomes and living standards in the two countries had converged in the 1890s, prompting Gladstone, the British prime minister, to refer to Britain and the United States as a "momentous" example "to mankind for the first time in history of free institutions on a gigantic scale."[11]

The speed with which the United States had transformed itself from a predominantly rural, agrarian society to a predominantly industrial, urban one, and had become the global symbol of economic success, could not help but astonish. When Alfred Marshall toured the country in 1875, farming and, to a lesser extent, mining were the principal sources of American income. By the time the Webbs visited, wages and profits from manufacturing had grown to three times those from agriculture. Between 1880 and 1900, the annual income generated by America's largest industries quadrupled. Income from printing and publishing in the United States jumped fivefold, machinery and malt whisky, fourfold; iron and steel and men's clothing, threefold. Electrification, refrigeration, new cigarette making, milling, distilling and other machinery, entirely new industries based on products derived from oil and coal, the extension of rail, and telegraph links to virtually every community produced a revolution in the scale, structure, and reach of American firms. Remington (1816), Singer (1851), Standard Oil (1870), Diamond Match (1881), and American Tobacco (1890) were born. The era of mass distribution, mass production, and scientific management—big business, in short—had arrived.[12]

Beatrice and Sidney were more interested in the machinery of American government than in the operations of American business. Their first stop was Washington, D.C., an unfortunate choice given that the capital was gripped with war fever. An uprising in Cuba, Spanish repression, and the sinking of the USS *Maine* in Havana Bay, blamed on Spain, had ignited a powerful grassroots movement in favor of military intervention. Prowar sentiment wore down opposition by business and religious leaders and the Republican president, William McKinley. Beatrice and Sidney sat with more than a thousand spectators in the visitors' gallery of the House of Representatives when President McKinley signaled his change of heart. Beatrice was appalled by the House and unimpressed by the Senate. She formed a more favorable opinion of Teddy Roosevelt, undersecretary of the navy, who was a leading proponent of war. She found his talk "deliciously racy" when he was telling stories of life on a Western ranch, though she was disappointed that he spent most of their lunch "breathing forth blood and thunder" and seemed utterly indifferent to local government, the subject of her next book with Sidney.[13]

New York did not suit Beatrice any better.

> Noise noise nothing but noise . . . In the city your senses are disturbed, your ears are deafened, your eyes are wearied by a constant rush; your nerves and muscles are shaken and rattled in the streetcars; you are never left for one minute alone on the road, whether you travel by ordinary car or Pullman; doors are opened and slammed, passengers jump up and down, boys with papers, sweets, fruits, drinks, stream in and out and insist on your looking at their wares or force you to repel them rudely; conductors shut and open windows; light and put out the gas; the engine bell rings constantly, and now and again the steam whistle (more like a fog-horn than a whistle) thunders out warning of the train's approach.[14]

She shared none of Marshall's or the Americans' love of technology and the mobility that modern technology implied. Not only the trains and skyscrapers but the "perfectly constructed telephones, skilled stenographers, express elevators, electric signals of all descriptions" left her cold. She was forced to acknowledge the "all-pervading and all-devouring 'executive' capacity of the American people" but attributed it to Americans' ready acceptance of "pecuniary self-interest as the one and only propelling motive." She quickly decided that a short attention span ("impatience") was the greatest flaw in the national character and was inclined to think that the speed of travel, of communication, and of American life in general—the "noise, confusion, rattle and bustle"—were merely so much wasted energy. "All this appreciation of mechanical contrivances seems to us a symptom of the American's disinclination to think beforehand," she commented.[15] She did not connect, as Marshall had, "nervous energy" with the zeal to manage, organize, and get things done or the love of risk with innovation or social mobility.

When Beatrice and Sidney went west a few weeks later, their first stop was in Pittsburgh. At Carnegie Steel, the "vast wealth-producing machine" that eventually became U.S. Steel, she was struck by the extent to which the technology had replaced labor. Henry Clay Frick gave Beatrice a tour of the Homestead, Pennsylvania, steel plant. He told her that Carnegie Steel had tripled its output while cutting the payroll from 3,400 to 3,000 in the space of a few years. She described

> acres of shops filled with the most powerful and newest machinery. The place seemed almost deserted by human beings. The great engines, cranes and furnaces were struggling and panting, seemingly without the aid of man. It was only now and again that one espied a man enclosed in a little cabin, swinging midway between the ground and the rafts of the shed, and working some kind of electrical machine whereby millions of horse power was set in motion and directed . . . We gathered that the great technological advances in labour-saving had been made in the past ten years, largely in the application of electric power to work new automatic machinery. The "traveling buggies" which replaced labour in moving the great masses of steel in and out of the rolling mills; the automatic machinery by which a single man swinging on a moving arm, opened the furnace door, lifted out the heated mass of steel, and swung it on to the buggy; and the automatic charging of the furnaces themselves with cradles of scrap steel also by a single man, were all introduced within the last six years.

She shrewdly attributed the phenomenal success of the Carnegie business less to "the mechanical contrivances"—which were accessible to steelmak-

ers anywhere in the world—than to superior management and organization. She noted that all the owners were working members of the privately
held firm that displayed "a lavishness towards all the brain-workers," who
were provided with "elegant homes . . . outings to Europe, and endless
treatings at home."[16]

The city, on the other hand, was

> a veritable Hell . . . which combines the smoke & dirt of the worst part
> of the Black country with the filthy drainage system of the most archaic
> Italian city. The people are a God-forsaken lot . . . tenements built back
> to back—crazy wooden structures crammed in between offices 20 stories
> high, streets narrow & crowded with electric trains rushing through at
> 20 miles an hour—altogether a most diabolic place with the corruptest
> of corrupt American governments.[17]

She saw what Charles Philip Trevelyan had warned her about before she
arrived: Andrew Carnegie, whom she called "the reptile," and other Pittsburgh tycoons may have "given a Park or two, a free library or so," but
had otherwise left the city "severely alone."[18] From Pittsburgh, Beatrice
rushed to Chicago, Denver, Salt Lake City, and San Francisco. By the time
she sailed to Hawaii on her way to New Zealand and Australia, she was
convinced that the rest of the world had very little to learn from America's
social experiment.

Before leaving New York, Beatrice had sought out a number of educa-
tors and economists. With the sole exception of Woodrow Wilson, later
the president of Princeton University, American academics impressed her
unfavorably. After a lunch at Columbia University, she compared one economics professor to "a superior elementary school teacher" and described
the campus as "something between a hospital and the London polytech-
nic." Yale was nothing more than "a pretty little conventional university."
About the economist who would become the author of the Sherman Anti-
trust Act, she grumbled that "from his appearance, manner and speech I
should have taken him for a pushing and enterprising manager of a store
in a Western city."[19]

. . .

Irving Fisher, the newest member of the economics faculty at Yale, was by no means mediocre or dull. His eyes sparkled with intelligence, his handshake was firm, he had the build and bearing of an athlete, and his face was boyishly handsome. At thirty years old, he was the only American economic theorist that Cambridge and the rest of England and the Continent took seriously. Alfred Marshall and Leon Walras, the French mathematical economist, considered him a genius.[20]

Named after Washington Irving, the author of "The Legend of Sleepy Hollow," Fisher was born in Saugerties, a farming community in New York's Hudson Valley, two years after the end of the Civil War. His grandfather was a farmer. His father, George Fisher, was a high-minded Evangelical minister. His mother, Ella Fisher, a former pupil of George Fisher's, was a strong-willed, devout young woman. When Irving was one year old, his father, who had recently graduated from Yale's divinity school, was offered a pulpit in Peace Dale, Rhode Island.

Peace Dale was a smaller, more picturesque version of Henry James's imaginary New England mill town in *The Ambassadors*. Like Woollett, Massachusetts, the town in which Irving Fisher spent his childhood was prosperous, paternalistic, and steeped in New England Evangelicalism. The town's leading citizen and benefactor was Rowland Hazard III, a Quaker, who had inherited the woolen mills his father had established and had founded a chemical company himself. Hazard was considered a progressive employer, having instituted profit sharing for his employees, and, on handing the reins of his business to his sons, he plunged into a second career as a political reformer. One of his daughters, Caroline, eventually became the president of Wellesley College. Hazard built the Congregational church and invited George Fisher to become its first pastor. Thanks to Hazard's patronage, Irving grew up in a rambling parsonage within sight of the Atlantic among "plain terms" and "honest minds."[21]

When Irving was thirteen, his father abruptly left his congregation and his family for a Wanderjahr in Europe, visiting great universities and cathedral towns. When he returned, a spirit of restlessness impelled him to take up temperance with great zeal, and he soon plunged his parish into bitter controversy. When his flock refused to support him, he resigned and moved his family into a cramped tenement in New Haven, Connecticut, where he enrolled Irving in a public school. For two years, the Fisher family relied on relatives for support.

When George Fisher finally found a new pulpit, it was 1,200 miles away on the Missouri-Kansas border. Missouri, Alfred Marshall wrote in 1875, was "full of swamps, negroes, Irishmen, agues, wildly luxuriant flowers & massive crops of . . . Indian corn" and St. Louis was a singularly "unhealthy town." [22] But neither heat nor humidity had deterred waves of migrants from the East who were attracted by rising wheat prices and appreciating farmland. Cameron, Missouri, was a jumble of rail yards, warehouses, feed lots, a few wide streets bordered by large houses, and at least one dozen churches. In the fall of 1883, when George Fisher left New Haven, he expected to send for his wife and younger son the following spring. Irving, now sixteen, went with his father as far as St. Louis, where he was to live with George Fisher's sister and brother-in-law, a professor at Washington University. Fisher had arranged for his son to finish high school at an elite Congregational preparatory school. His fervent hope was that his gifted older son would attend Yale and ultimately train for the ministry there.

When George Fisher continued his journey, father and son were separated for only the second time in their lives. They had expected to visit each other, but the distance between Cameron and St. Louis, three hundred or so miles, proved much too far in sleet, snow, and bitter cold. By the end of the first winter in Cameron, George Fisher was complaining of a strange lassitude, perpetual fever, and sinking spirits. These, it was quickly determined, were the classic symptoms of tuberculosis. In May, George, who was now very ill, made the long journey back east. He rejoined his wife and younger son in New Jersey at the home of a second brother-in-law, a physician who took the family in and cared for the dying man. Irving stayed behind. George Fisher insisted that he finish high school in St. Louis and take his college entrance examinations. By the time Irving graduated with honors, won a scholarship to Yale, and was reunited with his parents and younger brother in July 1884, his father was near death. He left behind a penniless widow, a ten-year-old, and seventeen-year-old Irving.

Fisher's grief was compounded by his disappointment at almost certainly having to postpone college, if not give it up altogether. The only prospect he could think of was to return to Missouri and look for a job on a farm that belonged to a classmate's family, where he had worked the previous summer.

A \$500 legacy that his father had invested with a friend in Peace Dale and designated for Irving's education was discovered. If Fisher lived with his mother and brother in a three-room apartment near Yale, his mother could rent out the second bedroom to another student and he could tutor. Together with his scholarship and the legacy from his father, this measure would just suffice to allow him to enroll at Yale as planned in the fall of 1884.

J. Willard Gibbs, one of Yale's "great men," had observed that if the masses were going to run the world, they would need a lot of instruction. Few pursuits in that era required a university education, and the sacrifice of four years' wages was beyond the means of all but 1 to 2 percent of young men. But by the 1880s, a growing number of small-town American boys "longing to escape from the inferiorities of childhood" began to view college as a promising exit strategy. In America's new industrial and urban economy, opportunities for engineers, accountants, lawyers, and teachers, not to mention managers in the new corporate concerns, were multiplying fast enough to become a means other than the usual "grind of moneymaking"—that path being a long, uncertain, and arduous affair—to achieve distinction.[23]

A poor but ambitious boy like Fisher was lucky that family money was so commonplace an asset among Yale undergraduates that its social value was greatly depreciated. Popularity and fame required prowess as an athlete, an orator, a debater, a wit, or even a scholar. Fisher rowed for college, dazzled the faculty in the Junior Exhibition debating competition, won coveted prizes in mathematics and other subjects, and graduated first in his class of 124.[24] But the acme of his college career came the day he was tapped to become a member of the elite secret society Skull and Bones.

Muriel Rukeyser, the poet, observes in her biography of Gibbs that this was "the season for the young sciences" in America.[25] The 1880s saw an explosion of scientific activity in the United States and rising popular interest in science. Charles Darwin, Herbert Spencer, and Alfred Russel Wallace, the independent discoverer of evolution by natural selection, became household names; zoos and natural history museums proliferated; and the science-fiction novel was born. Edward Bellamy's *Looking Backward: 2000–1887*, which fast-forwarded readers to Boston in the year 2000, depicted a golden age of phonographs, credit cards, and radios.[26] New professional societies, scientific publications, and laboratories were popping up like mushrooms after a rain, while universities shifted their focus from training young men in the classics to turning out scientific and technical workers. The Brooklyn Bridge, which opened in Fisher's senior year in high school, symbolized the power of science to transform society. The rise of huge enterprises and entrepreneurial fortunes and the role of railroads in economic growth stimulated interest in finding new "instruments of mastery."[27] In the popular imagination, science was increasingly being seen as a way to get rich and, at the same time, a vehicle for solving the myriad social ills of poverty, disease, and ignorance.

Gibbs was a physicist, chemist, and mathematician, who was the first to apply the second law of thermodynamics to chemistry. The function of the scientist, he once said, was "to find the point of view from which the subject appears in its greatest simplicity."[28] He was a great champion of the mathematization of science. Mathematics was a lingua franca as well as a tool of analysis, so mathematics could promote the global exchange of ideas among scientists, just as Latin had done for centuries among botanists and anatomists. Gibbs almost never spoke at faculty meetings. But at the end of a fractious debate over whether mathematics could substitute for Greek or Latin to fulfill Yale's classical language requirement, he rose, coughed politely, and was heard to murmur as he left the room: "Mathematics *is* a language."[29]

By the time Fisher was a senior, he considered himself a mathematician but longed for more. "I want to know the truth about philosophy and religion." [30] He had rejected the idea of becoming a minister like his best friend from St. Louis, Will Eliot. At different times, he thought of the law, railways, public service, and science. "How much there is I want to do! I always feel that I haven't time to accomplish what I wish. I want to read much," he wrote in a letter to Eliot. "I want to write a great deal. I want to make money." [31] Ultimately, Fisher chose the "science of wealth."

American economics of the Progressive Era is typically described as utterly divorced from the British evolution toward collectivism and the welfare state. Except for a few so-called institutionalists such as Thorstein Veblen who were critical of commercial society, academic economics is supposed to have been dominated by Social Darwinists who defended laissez-faire and the rich and wanted to trample the poor.

It simply wasn't so. Virtually every founding member of the American Economic Association got his training and worldview at Berlin, Göttingen, or Munich and shared the values of the German "historical school," which, in opposition to English economics, explicitly condemned unfettered competition and championed the welfare state. Arthur Hadley, who held a chair in political economy at Yale, once referred snidely to American economists as "a large and influential body of men who are engaged in extending the functions of government." [32] The economics department at Yale was no exception, but for its most notorious member, William Graham Sumner. Warning that contemporary political labels— conservative versus liberal, left versus right—fit nineteenth-century intellectuals poorly if at all, the historian Richard Hofstadter once asked rhetorically about Sumner "whether, in the entire history of thought, there was ever a conservatism so utterly progressive?" [33] The son of an English immigrant laborer and an Episcopalian minister, Sumner was both a political economist and America's first sociologist. Austere and satirical with clipped, grizzled hair, Sumner taught himself "two Scandinavian tongues, Dutch, Spanish, Portuguese, Italian, Russian, and Polish" when he was in his late forties and turned "New Haven into a kind of Social-Darwinian pulpit" for his libertarian views. His lectures were described by contemporaries as "dogmatic," his manner "frigid," and his voice like "iron." [34] But his passionate delivery and fearless espousal of contrarian views made him the most popular lecturer at Yale.

Sumner was a great admirer of Charles Darwin and Herbert Spencer. He objected not only to bigger government but also to the activities of most private charities. His economics were thoroughly Malthusian, that is to say, deeply pessimistic, and, like Malthus, Ricardo, and Mill, he dismissed all schemes to speed up society's evolution as snake oil, stupidity, special pleading, or "jobbery." Yet, like the economic thinkers he admired, he was by no means a defender of the status quo. Trained as a minister, Sumner was as likely to condemn war as welfare, to defend striking unions as well as the banker Andrew Mellon's right to his millions, and to praise working women in the same breath as free trade. When Yale's president tried to forbid, on theological grounds, his use of Spencer's *Principles of Sociology* as a text, Sumner threatened to resign. Around the time of the Webbs' visit, Sumner enraged Yale's Republican alumni into calling for his dismissal, by publicly condemning the Monroe Doctrine and the Spanish-American War.

According to his son, Irving Norton Fisher, Fisher signed up for every course Sumner gave. He approached economics as a mathematician or an experimental scientist, describing himself to Eliot at one point as "your cold analytical mathematical friend." [35] Not long after Sumner introduced him to the subject, he concluded that a great deal might be accomplished by someone trained to think like a scientist—that is, coldly, analytically, mathematically.

When Fisher consulted Sumner about a dissertation topic in the spring of 1890, the latter's personal interests were shifting away from classical political economy to "the science of society." Already deep into his extraordinary spurt of late-life language acquisition, and busy collecting ethnographic data, Sumner wished to put sociology on a more rigorous footing. In this spirit, he suggested that Fisher write his doctoral thesis on mathematical economics, a subject that was both new and beyond the technical capability of most older economists, including himself. He lent

Fisher a volume by William Stanley Jevons, one of the pioneers of a new method involving calculus for analyzing consumer choice by focusing on marginal changes.

The impulse to make their respective fields more scientific was spurring ambitious young humanists to seize scientific knowledge as their special tool. The psychologist and philosopher William James, just back from Europe, wrote to a friend that year, "It seems to me that perhaps the time has come for psychology to begin to be a science." [36] Fisher already considered mathematics an ideal global currency that encouraged trade in ideas. He was intrigued by the prospect of strengthening the theoretical foundation of political economy, as Gibbs had done for that of chemistry:

> Before an engineer is fit to build the Brooklyn Bridge or to pronounce
> on it after it is built it is necessary to study mathematics, mechanics, the
> *theory* of stress and of the natural curve of a hanging rope, etc., etc. So
> also before applying political economy to railway rates, to the problems
> of trusts, to the explanation of some current crisis, it is best to develop
> the *theory* of political economy in general. [37]

Crude social Darwinists and their Socialist opponents identified competition as the distinctive feature of the modern economy, comparing the operation of markets to the laws of the jungle. But, like Marshall, Fisher was more impressed by the high degree of interdependence and cooperation among economic agents—households, firms, governments—and the large number of channels through which a given cause produced its ultimate effect.

Fisher occasionally went into New York City from New Haven, and on several occasions he visited the stock exchange. The operations of the market for securities were very much on his mind when he read the books Sumner had given him. He was struck that economists had evidently borrowed much of their vocabulary from the older science of physics; they spoke of "forces," "flows," "inflations," "expansions," and "contractions." But, as far as he knew, no one had actually tried to construct an actual model of the process that resulted in "that beautiful and intricate equilibrium which manifests itself on the exchanges of a great city but of which the causes and effects lie far outside." [38]

Marshall had conceived of modern economics as an "engine of analysis" and used graphs to trace the effects of external influences on individual markets. Fisher decided to construct a mathematical model of an entire economy. He wanted to be able to trace how a market "calculated" the prices that equated supply and demand. A practical Yankee, he wanted a model that could spit out numerical solutions, not just mathematical symbols. Almost as soon as he started working on his model, Fisher decided to take his project a step further and build a physical analog of the equations in the form of a hydraulic machine. That is something that probably would only have occurred to a tinkerer who had spent hundreds of hours in a laboratory performing tedious and repetitive physical experiments. Fisher asked Gibbs, who was far more able than Sumner to grasp what he was trying to accomplish, to read his manuscript.

In Fisher's model, everything depends on everything else. How much of a commodity each consumer wants depends on how much of every other commodity he wants. Fisher acknowledged that the bulky contraption, with its cisterns, valves, levers, balances, and cams applied "imperfectly at best" to the exchanges of "New York or Chicago" but he was in no way apologetic. "Ideal suppositions are unavoidable in any science," wrote the doctoral candidate in his thesis. "The physicist has never fully explained a single fact in the universe. He approximates only. The economist cannot hope to do better." [39]

The marvelous physical contraption let someone visualize the elements whose interplay produced prices. It also permitted someone to "employ the mechanism as an instrument of investigation" of complicated and distant interconnections. For example, one could see how an external shock to demand or supply in one market affected all the prices and amounts produced in ten interrelated markets, altered prices and quantities in all markets, and changed the incomes and choice of products purchased by various consumers. Fisher's hydraulic machine was the precursor of the simulation and forecasting models with thousands of equations that were developed in the 1960s to run on huge mainframe computers and that today's undergraduates can use to calculate a country's GDP on a notebook computer. Alas, both Fisher's original model and a replacement constructed in 1925 when the first was destroyed en route to an exhibition have been lost to posterity.

Fisher wrote his thesis over the summer of 1890. He showed his enthusiasm for mathematical methods by including an exhaustive survey and bibliography of applications. The economist Paul Samuelson called "Mathematical Investigations in the Theory of Value and Prices" "the greatest doctoral dissertation in economics ever written."[40] When it was published, the *Economic Journal,* founded by Alfred Marshall and other members of the newly inaugurated British Economic Association, greeted it as a work of genius. The reviewer, Francis Ysidro Edgeworth, an Oxford professor and one of the founders of mathematical economics, wrote, "We may at least predict to Dr. Fisher the degree of immortality which belongs to one who has deepened the foundations of the pure theory of Economics."[41] In the third edition of his *Principles,* Marshall, who could be stingy when acknowledging the contributions of other scholars, included not one but three highly flattering references to Fisher's "Investigations," referring to it as "brilliant," classing Fisher with "some of the profoundest thinkers in Germany and England."[42]

Fisher's picture of economic reality—especially his awareness of interdependence and mutual causation—affected how he thought on a great many other subjects. Just before he received his doctorate, he read a paper to the Yale Political Science Club proposing an international body representing all the nations of the world and dedicated to the peaceful settlement of international conflicts. According to the historian Barbara Tuchman, this paper later inspired the formation of the League to Enforce Peace, which, in turn, is credited with stimulating President Wilson's interest in forming a League of Nations.[43]

By the early 1890s in America, the post–Civil War railway, mining, and land boom had stalled, exposing the shaky nature of much of the finance. The panic of 1893 and the collapse of the stock market was followed by the worst depression in American history to date. Fisher's letters to his friend Will Eliot are as free of any mention of these calamities as Jane Austen's novels are of references to the Napoleonic Wars. Possibly his reason was similar to the author's: his mind was on love, courtship, and marriage.

Characteristically, Fisher deferred a return to Peace Dale, the scene of childhood happiness, until he could ride into town decked in a hero's laurels. When he left, he had been thirteen and deeply unhappy. When he returned, he was wrapped in the cloak of a "brilliant career at Yale, as prizeman, valedictorian, instructor and now professor of mathematics."[44] Like the hero of a Victorian three-decker novel, his object was to claim the heiress—or, since this was America, the boss's daughter. The way it happened was purely providential. It took little more than a glance for Irving to fall in love with a former childhood playmate; Margaret Hazard, or Margie, as she was called.

Margie Hazard was blessed with a sheltered upbringing, a serene temperament, and an unusually sweet nature. Her sister was the intellectual, Margie the creative, maternal one. Her faith in Irving was complete and unwavering. She was a wealthy heiress while he was penniless, yet she was convinced that she was the luckiest of women. They married in June 1893 with the entire population of Peace Dale there to witness the ceremony and take part in the festivities. The vows were read by no fewer than three ministers, and the wedding cake weighed fifty pounds. At a time when every day brought news of a fresh bankruptcy or bank run, some were scandalized by this display of conspicuous consumption. So it was just as well that the bridegroom and bride slipped away to New York, boarded an ocean liner, and sailed to Europe for a yearlong honeymoon.[45]

"All educated Americans, first or last, go to Europe," Ralph Waldo Emerson noted sourly. The rich went on the obligatory "grand tour" of capitals; the intellectually ambitious, on a "grand tour" of universities.[46] Zigzagging across England and the Continent by rail in 1893 and 1894, Fisher found it possible to exchange ideas with virtually every prominent member of the small, if growing, economics fraternity. His "little book . . . made a small path" for him through Europe, giving him instant entree to a cosmopolitan fraternity of economic thinkers. Fisher lunched in Vienna with Carl Menger, the founder of Austrian economics. He dined in Lausanne, Switzerland, with Leon Walras. Walras's brilliant student Vilfredo Pareto joined them, and Pareto's wife shocked Fisher by lighting up at tea. He stopped in Oxford to confer with the tongue-tied and absentminded Ysidro Edgeworth, and made a pilgrimage to Cambridge to pay his respects to Alfred Marshall, whose recently published *Principles* had cemented his stature as the world leader of theoretical economics.

Despite the hectic pace of his travels, Fisher still had plenty of time to attend the lectures of the mathematician Henri Poincaré in Paris and the German physicist Hermann Ludwig von Helmholtz in Berlin. When northern Europe got too cold for his now pregnant bride, he hired another student to take notes for him and took her to the French Riviera. Hiking in the Alps alone, he experienced an epiphany as he watched water tumble over rocks and gather in a pool below. "It suddenly occurred to me that while looking at a watering trough with its in-flow and out-flow, that the basic distinction needed to differentiate capital and income was substantially the same as the distinction between the water in that trough and the flow out of it."[47] After Fisher gave a talk at Oxford, Edgeworth told Margie, who had joined her husband, "Professor Fisher is soaring."[48]

By the time Fisher and his wife returned to New Haven and a brand-new, fully furnished mansion, thoughtfully provided by the Hazards, the mood of the nation was grim. By 1895, more than five hundred banks had failed. Fifteen thousand companies had declared bankruptcy. Unemployment had idled one in seven workers.[49] The fiery blast furnaces and behemoth textile mills were still standing, the vast railroads were still capable of hauling freight, and the prairies were still golden with wheat and corn. Yet amid this potential feast, there was a kind of famine. "Never within my memory have so many people literally starved to death as in the past few months," a Reverend T. De Witt Talmage told his congregation. "Have you noticed in the newspapers how many men and women here and there have been found dead, the post mortem examination stating that the cause of the death was hunger?"[50]

Popular anger against "money men" was ubiquitous. James J. Hill, the founder of the Great Northern Railway, wrote to a friend that "lately the people of the country are fixing their minds on social questions . . . For ten years it has been 'railroads, monopolies and trusts' but now it shows up as those who have nothing against those who have something."[51] That year, a melodrama by Charles T. Dazey called *The War of Wealth* opened on Broadway.

The depression aggravated longstanding social and political conflicts. These were not primarily between classes: the 1894 Pullman strike notwithstanding, the number of strikes had drifted down each year. Rather, they were clashes between regions, between the representatives of different industries, between small and big business. Western silver miners blamed the collapse of metal prices on Washington. Farmers blamed their debt troubles on voracious eastern bankers and pitiless railroad monopolies. Of all constituencies, they were the angriest. The boom had passed them by, and the bust was driving them to despair. Amid a general slide in prices, those of wheat and corn and sugar had plunged two and three times more, on average, than other prices. Everyone connected with agriculture was drowning in debt, oppressed by high interest rates, and terrified of foreclosure.

The presidential campaign of 1896 became a referendum on the country's economic direction. The Democratic incumbent, Grover Cleveland, was repudiated by his own party. William Jennings Bryan, the thirty-six-year-old Democratic nominee, promised his western constituents that he would "nationalize the railroads, sweep away the tariff, and, most of all, rid them of financial tyranny." He called eastern bankers "the most merciless unscrupulous gang of speculators on earth," the "money monopoly."[52] His critics returned the compliment by calling him an anarchist, a Benedict Arnold, an Antichrist, "a mouthing, slobbering demagogue."[53] His Republican opponent, handpicked by James J. Hill and other tycoons, was William McKinley.

Six weeks before the election, Bryan had already nailed Wall Street onto his cross of gold when he took his presidential campaign up to the doors of one of the Jerichos of money power. At Yale University on the first day of the fall semester, the "Great Commoner" faced one thousand undergraduates and professors. Wild boos and cheers erupted as soon as the handsome bear of a man with flowing dark hair wearing a black felt hat and a string tie mounted the platform.

"The great paramount issue" of the 1896 election, he told them, was the seemingly obscure question of the nation's monetary standard. In a deep, slightly hoarse voice, Bryan inveighed against "a gold standard which starves everybody except the money changer and the money owner." The embrace of gold in the 1873 act banning the free coinage of silver had produced a money drought that he said was far more devastating to the nation's biggest industry, agriculture, than any act of nature. "If you make money scarce you make money dear," Bryan told the crowd. "If you make money dear you drive down the value of everything and when you have falling prices you have hard times." [54]

According to Bryan, the only way to revive the economy was to make money cheap again, that is, by tying the dollar to a more expansive standard than gold "that permits the nation to grow." He accused McKinley, and the "Gold Democrats" who supported him, of perversely trying to restore prosperity by continuing the disastrous "sound money" policies of the Democratic incumbent. In the fourth year of the depression, McKinley and the sound money clubs his supporters had organized were more worried about inflation and the London money market than the suffering at home. What was bad for the farmer was bad for America, including its small businessmen, professionals, and factory workers—and the students of New Haven. If the silver standard could ruin businessmen "with more rapidity than the gold standard has ruined them, my friends, it will be bad indeed," Bryan told the crowd, adding that the political "party that declares for a gold standard in substance declares for a continuation of hard times." [55]

At the mention of the Republican Party, the students began yelling, jeering, and bellowing McKinley's name. Uncharacteristically, Bryan lost his temper: "I have been so used to talking to young men who earn their own living," he shouted, "I hardly know what language to use to address myself to those who desire to be known, not as creators of wealth, but as the distributers of wealth which somebody else created."[56] A sophomore later recalled Bryan's next words, later denied by the candidate: "Ninety-nine out of a hundred students in this university are sons of the idle rich." The word *ninety-nine* had the effect of a starter gun at a race. "Ninety-Nine! Nine, Nine, Ninety-Nine!" the class of '99 chanted until Bryan abandoned the stage in disgust, leaving the money changers still in possession of their temple.[57] The *New York Times* crowed the next day: "YALE WOULD NOT LISTEN; Derisive Applause and a Brass Band Too Much for the Boy Orator—He Spoke Only About Twenty Minutes and Retired in Disgust."[58]

"I was never so *morally* aroused, I think, as against the 'silver craze,'" Irving Fisher confided to his friend Will Eliot in a letter.[59] "Social science is very immature and . . . it will be a very long time before it reaches the "therapeutic stage."[60]

Fisher had recently switched from Yale's Department of Mathematics to its Department of Political Economy, largely out of a desire to "make direct contact with the living age," although he privately was of the opinion that its members were "eaten with conceit" and overly confident that they knew how to fix the world's ills. He was as trim and upright as ever, keeping in shape with a regular regimen of jogging, rowing, and swimming, and was unflaggingly energetic. The only mark of time's passage was blindness in his left eye, the unhappy consequence of a squash accident.[61]

Fisher had few strong political convictions but found that, as a professor, "I am expected to have an opinion."[62] Misguided reform was likely to make matters worse, he warned. Sumner had expressed profound misgivings about populist measures in a pamphlet provocatively titled *The Absurd Effort to Make the World Over*.[63] In the depression that followed the panic of 1893, Fisher had written to his friend Will:

Concerning social reform, I feel that the effort of philanthropists to apply therapeutics too soon is more likely to lead to evil than good. The very best the exhorter can do is to work *against* the "something must be done" spirit, and beg us to wait patiently until we know enough to base action upon and meantime confine philanthropic endeavor to the narrow limits in which it has been proved successful—chiefly education . . . There is so much *specific* reform at hand to be done—in city government, suppression of vice, education—that the hard workers of humanity need not and ought not talk, until "little" things are done, on broad schemes for "society."[64]

As it turned out, Fisher did not follow his own advice. At a meeting of the American Economic Association in November 1895, he was scandalized at the "too lighthearted way" that some of his colleagues were willing to "tamper with the currency" and delivered a stinging critique of the silverite argument. "The effect of bimetallism, if silver is the cheaper metal, must be a depreciation of the currency. . . . A system whose claim to recognition is based on considerations of justice has no excuse when beginning with such a glaring injustice. Honest men must regard with horror the proposal to reintroduce a ratio of 15½ to 1." Not surprisingly, the speech attracted the gratified attention of the anti-Bryan forces. Fisher let himself be recruited to the New York Reform Club's sound money committee and the anti-Bryan campaign.[65]

How money became the paramount issue of the 1896 presidential campaign requires some explanation. Historically, money had been seen as powerful, desirable, very likely evil, and mysterious, like natural calamities or epidemics. Interest was traditionally treated with hostility by Christianity as well as by Islam. Financial crises—from stock market crashes, to bank runs, to hyperinflations—sparked popular rage against bankers. The subject was shrouded in myth, superstition, and emotion.

In the 1880s and 1890s, both sides in the populist debate mythologized their metal of choice and demonized their opponents. The evil speculator became a stock figure of fiction in the 1880s, his way paved by the Nibelungs in Richard Wagner's opera *The Ring of the Nibelungs* and Auguste Melmotte in Anthony Trollope's novel *The Way We Live Now*. The historian Harold James comments:

The stories that the nineteenth century told about the global world built
on a secular concept of original sin. The remedy that many thinkers then
provided to the illegitimacy of the system echoed Luther's quite precisely
(in a secular manner). Strong public authority was needed to overcome
the legacy of that sin. There was a natural community that had been broken apart by creative greed, but the state could create its own order and
community, and thus channel the destructive forces of dynamic capitalism. This strategy would offer the only way of avoiding the apocalyptic
crisis prophesied by a Marx or Wagner or a Lord Salisbury."[66]

American economists were always more obsessed by "the money question" than their English counterparts were. But this was largely an accident
of history, resulting partly from long-standing American suspicion of fed-
eral power and partly from the decision to issue unconvertible greenbacks
during the Civil War and to allow them to be redeemed for gold twenty
years later. More compelling, bank runs, financial panics, crises, and depressions occurred with dangerous frequency. The English financial writer
Bagehot had observed in 1873:

> It is of great importance to point out that our industrial organisation
> is liable not only to irregular external accidents, but likewise to regular
> internal changes; that these changes make our credit system much more
> delicate at some times than at others; and that it is the recurrence of
> these periodical seasons of delicacy which has given rise to the notion
> that panics come according to a fixed rule—that every ten years or so we
> must have one of them.[67]

In the face of such traditional fatalism, it seems entirely plausible that an
idealistic young scientist protested that the real problem was that money
had not been sufficiently or rigorously studied and that a better understanding of money's role in economic affairs would minimize irrational
decisions and unnecessary conflicts.

In his PhD thesis, which had been published in 1892, Fisher commented that "money, which is used to measure value and therefore affects

all perception of economic values, is little studied and the mystery that surrounds money is at the root of many misunderstandings and miscalculations." Although the focus in that investigation was how prices were "computed" through the interaction of supply and demand, Fisher treated money first and foremost as a unit of measurement. The gold standard was a primitive mechanism for tying down its value. But even as he wrote his thesis, he developed a potentially better way. He saw that it might be possible to stabilize prices by tying the dollar's value in terms of gold to an index of consumer prices. Fisher saw equilibrium as a reference point and monetary disturbances as the source of instability. In *Mathematical Investigations,* he stressed that "the ideal statical condition assumed in our analysis is *never* satisfied in fact," which convinced him that "panics show a lack of equilibrium."[68]

Interest is the price that those with savings charge to let others use their capital, a real and valuable service. The value of capital, in turn, is determined by expectations on the part of savers and investors about the future stream of interest payments. Inflation and deflation produce large and arbitrary shifts in income and are the effects of the fluctuating value of the monetary standard—a rubber yardstick rather than a constant one—not conspiracies by demagogues and mobs on the one hand or Wall Street bankers on the other.

Having come to economics via the monetary debates that dominated American politics in the 1880s and 1890s, Fisher was concerned primarily with justice for debtors and lenders and with avoiding social conflicts that were exacerbated by unexpected changes in the value of money. As a practical matter, it was difficult for an individual businessman to distinguish between a change in the price of his product and an overall rise or fall in prices, and to adjust his contracts accordingly. Citizens who didn't understand that the value of their currency wasn't fixed tended to blame scapegoats—easterners, Jews, foreigners—for inflation or deflation.

The United States had followed Britain, Germany, and France by adopting the gold standard—a system in which each country's currency is pegged to a fixed amount of gold and hence to fixed amounts of other currencies. Think of it as a single world currency, the existence of which is a great convenience to those engaged in exporting and importing. Kansas farmers who sold wheat to British merchants wanted dollars with which to pay the men, railroads, seed suppliers, and so forth. So the British merchants were obliged to buy dollars with pounds. Obviously, knowing that £1 can always be exchanged for $5 is the next best thing, from the point of view of the importer, to a single currency.

Unfortunately, fixing exchange rates did not mean, as many supposed, that the value of the currency in terms of domestic goods was constant. Indeed, while the United States pegged the dollar to a certain amount of gold, the gold's and therefore the dollar's domestic purchasing power fluctuated by as much as 50 or 100 percent. In the 1880s the dollar's value rose sharply as a result of a worldwide shortage of gold, producing price deflation and a raging debate between those who wanted to remain on the gold standard and those who wanted to return to a silver standard.

American farmers, who tended to speculate in land and to use mortgages to finance land purchases, were net debtors. They argued that maintaining gold parity had restricted the supply of money and caused interest rates to rise and crop prices and farm income to fall. That meant that more tons of corn or wheat or bales of cotton were needed to pay off or service a given amount of debt than the farmer or bank had anticipated when the mortgage was issued. Fisher had some firsthand knowledge of western farming, thanks to his friendships with the sons of Missouri farmers during his two years in St. Louis and his summer jobs on their farms during college.

The free-silver movement reached its apogee in William Jennings Bryan's presidential campaign of 1896, and so did Fisher's defense of the gold standard. His monograph *Appreciation and Interest* had just been published. For him, the issue was distributive justice. Fisher conceded that the "silverites" were right in claiming that deflation had enriched lenders at the expense of debtors. But the argument for switching to a silver standard was faulty all the same. In fact, he claimed, declining interest rates automatically offset the rise in the real value of their debt. The market compensated . . . Bryan lost the election. Ironically, right around the time of his "Cross of Gold" speech, gold discoveries and other developments produced a spurt in the gold supply and led to a monetary expansion that brought the deflation of the 1880s and 1890s to an end without the United States abandoning the gold standard.

At thirty, Irving Fisher was the author of several books and monographs, a rising factor in the academic world, and the father of a growing family. He was stronger, handsomer, and more energetic than at twenty. He cycled, walked, and lifted weights. His favorite sport was swimming, and he let nothing—not even the chilly water off Maine or Margaret's anxiety—keep him out of the water in the summer.

In August 1899, Fisher was swimming off the family's summer estate when he nearly drowned. In the weeks that followed, he developed lassitude, a low fever, and a deepening depression, symptoms ominously reminiscent of the initial signs of his father's deadly illness. Shortly after his thirty-first birthday and his promotion to full professor, he received a death sentence in the form of a diagnosis of tuberculosis.

Tuberculosis was the nineteenth-century AIDS, writes historian Katherine Ott. At the turn of the twentieth century, one in three deaths in major cities was due to consumption, and most of the victims of the "white plague" were young adults. The course of the illness was ghastly, and recovery rates were depressingly low. Victims dreaded the loss of work and ostracism that inevitably followed a positive diagnosis. One man wrote that when the doctor told him it was tuberculosis, the words "might just as well have been followed by 'The Lord have mercy on your soul,'" for he felt himself a dead man.[69] Fisher remembered his dying father, shrunken and skeletal, totally deaf, unable to swallow anything but driblets of milk and barely capable of speech. George Fisher had lingered in this state for several agonizing weeks. When he died he was just fifty-three.

Most treatments involved rest, fresh air, and a rich diet. The "mind cure" blamed the disease on the stresses of modern life and coincided with a craze for all things Japanese or Chinese. Its practitioners urged individuals to take responsibility for their own health and advocated "calming one's fractious thoughts so that one might connect with the powerful and invisible spirit of God, humanity or some other force."[70] This was the era of positive thinking. In a speech at a local boys' school, Fisher explained his personal philosophy:

All greatness in this world consists largely of mental self control. Napoleon compared his mind to a chest of drawers. He pulled one out, examined its contents, shut it up, and pulled out another. Mr. Pierpont Morgan is said to have a similar control. . . . What we call the *life* of a man consists simply of the stream of consciousness, of the succession of images which he allows to come before his mind. . . . It is in our power to so direct and choose our stream of consciousness as to form our character into whatever we desire.[71]

For the next six years, Fisher struggled to recover his health, natural energy, and normal high spirits. He spent nearly half a year at the Adirondack Cottage Sanitarium in Saranac, New York. The hospital was operated by Dr. Edward L. Trudeau and modeled on the Alpine sanitaria described by the German novelist Thomas Mann in *The Magic Mountain*. The children were dispatched to their grandparents, and Margie accompanied Fisher to Saranac. They bought a raccoon coat and a copy of John Greenleaf Whittier's long poem *Snow-Bound* to read aloud. "The doctors fully expect me to get well but it takes time," Fisher wrote to Will Eliot in December 1898. "I am sitting out on the porch, the thermometer is twenty and the snow is two feet deep. I find ink freezes and so use pencil."[72] By January 1901, Fisher's doctors told him that he was fully recovered, but it took him three more years to regain his former energy.

Surviving tuberculosis awakened the latent preacher in Fisher. He became a crusader for public health and an advocate for healthy living and mind control, to which he believed he owed his recovery. His triumph over tuberculosis convinced Fisher that the extraordinary—such as the doubling of average life spans by the year 2000—was possible. When he met Dr. John Harvey Kellogg, the crusader for "biologic living," Fisher told him that he was "on a quest not like Ponce de León for the fountain of youth but for ideas which may help us to lengthen and to enjoy youth."[73] Influ-

enced by Kellogg, Fisher conducted experiments on vegetarian diets with
Yale athletes as subjects, applied for the job of secretary of the Smithson-
ian, and lobbied for the creation of a cabinet-level health department. In
1908, after the assassination of President McKinley, Fisher was appointed
by his successor, Theodore Roosevelt, the youngest president of the United
States, to the National Conservation Commission. The idea of conservation "has its center of gravity in our sense of obligation to posterity." It is
hard for us in America, he noted, enjoying the present plenty, to realize
that "we are scattering the substance that belongs to future generations."[74]

In 1906, the year of the San Francisco earthquake, Fisher declared Homo
economicus, economic man, defunct and laissez-faire a dead ideology. At
a plenary address at the American Association for the Advancement of
Science, he called the acceptance of government regulation and welfare
measures "the most remarkable change which economic opinion has un-
dergone during the last fifty years."[75] Experience, he said, proved that the
basic tenets of liberal theory—that individuals were the best judges of self-
interest and that the pursuit of self-interest would produce the maximum
good for society—were wrong. Government regulation and voluntary reform movements—the nineteenth-century equivalent of today's NGOs—
were not merely not harmful, but necessary. Indeed, he said, they had
already done much to preserve the natural environment and to improve
public health. He said that if he had to choose between Sumner's extreme
libertarianism or Socialism, he would choose the latter, and he enumer-
ated many instances in which what is good for the individual is not good
for society and, therefore, laissez-faire is not the right policy.

The Nature of Capital and Income, published in 1906, reflected
Fisher's growing understanding of capital as a stream of future services and interest in conservation. Fisher was convinced that economic
interdependence—epitomized by urbanization, economic specialization,
and globalization—implied a greater need for data, education, coor-
dination, and intervention on the part of the government. He argued
that concern for the future required prevention and conservation. His
near-death experience gave his concern for economic efficiency and

prevention of waste even greater urgency. Perry Mehrling, a historian of economic thought, says that Fisher was influenced by a contemporary of Adam Smith, John Rae, to define "interest"—including profits, rents, and wages—as the value of the stream of services from the machines, land, and human capital accumulated in the past. All of Fisher's reforms, observes Mehrling, from increasing life spans to preventing depressions and wars, were directed at increasing current national wealth.[76]

Today, economists talk of bounded rationality, externalities, and market failures. Fisher spoke of ignorance and lack of self-control. More radically, he argued that even when individuals behaved totally rationally, the combined effect of their actions might reduce collective well-being. "Not only is it false that men, when let alone, will always follow their best interests, but it is false that when they do, they will always thereby best serve society."[77] One special kind of ignorance, he explained, involved treating the present as if it was the norm. Life spans were only half as long as they could be, he thought. Productivity was just half as great. His most intriguing insight was that the mind plays tricks. He called it the money illusion. For Fisher, inflation and deflation—all changes in the overall level of prices—were evils because they misled people into making bad decisions. At the level of the economy, the money illusion meant that it took a long time for businesses and consumers to adjust to changes in prices and interest rates.

He drew two conclusions from the recognition that Homo sapiens were not Homo economicus, hyperrational calculating machines. First, there was a strong case for compulsory education. Second, there was an even stronger case for regulating individual behavior, whether via fire regulations in tenements or the prohibition of gambling or alcohol and other drugs: "It is not true that ignorant parents are justified in imposing their ideas of education upon their children; hence the problem of child-labor, instead of concerning only the individual, as was at one time thought, has important and far-reaching relations to society as a whole."[78]

Fisher went much further than Marshall in pointing out the limits of the competitive model. In this he anticipated the entire arc of economic theory after World War II. "Even when government intervention is impracticable or inadvisable, there will still be good reason for attempting betterment of conditions through the influence of one class upon another, hence come social agitations."[79]

Even if everyone were perfectly rational, the pursuit of self-interest wouldn't necessarily add up to socially desirable results. "Individual action would never give rise to a system of city parks, or even to any useful system of streets," he said. Hence, he rejected privatizing either the money supply, as Spencer advocated, or the "still more astonishing suggestion [is] that even the police function of government should be left to private hand, that police corps should be simply voluntary vigilance committees, somewhat like the old-fashioned fire companies, and that rivalry between these companies would secure better service than now obtained through government police!"[80]

Fisher's illness was followed by an extraordinary burst of creativity. In the space of five or six years, he poured out ideas that had germinated during his enforced exile, in which he had embraced Indian philosophy and meditation practices.

> Last night at sunset I sat out there like an Indian, thinking of nothing, but *feeling* the serenity and power of the Universe ... Those subconscious impressions of three years or more of depression, fear and worry are still in my mental storehouse, but buried, I hope permanently. It has been only by hard work and the application of auto-suggestion that the blue devils have been crowded down at all. I have to confess that the chief thing the matter with me after the first year was fear ... Optimism is not a question of what evil exists nor of what we may expect of the future. A man may believe the world unhappy and that the earth will grow cold and dead, that he himself is to have pain, loss of friends, honor, wealth—and yet be an optimist.[81]

The year 1907 was an unsettled one in the financial markets. Fisher was hurrying to finish a new book, *The Rate of Interest*, that he subtitled "Its Nature, Determination and Relation to Economic Phenomena."

For the first time, he explicitly couched his theory in terms of a lack of foresight: periods of speculation and depression are the result of inequality of foresight. "A panic is always the result of unforeseen conditions and among those unforeseen conditions and partly as a result of other unforeseen conditions is scarcity of money on loan."[82]

If inflation or deflation were correctly anticipated, explains Perry Mehrling, interest rates in money markets would adjust instantly and perfectly. If lenders expected the overall price level to rise, they would require borrowers to pay a commensurately higher interest rate. If they expected the overall price level to fall, they would be willing to accept a commensurately lower interest rate. By the same reasoning, if borrowers expected higher inflation, they would realize that paying nominally higher interest rates would not affect their real rate of return. If borrowers expected deflation, they would realize that they could afford to pay only a commensurately lower nominal interest rate. In short, if anticipations were correct, changes in the price level would have no effect on real output or employment. The trouble, of course, is that such perfect foresight is impossible: "Their failure to [correctly anticipate deflation] results in an unexpected loss to the debtor and an unexpected gain to the creditor."[83]

Now Fisher reversed his earlier position that changes in the value of money have a negligible effect in real economic activity and decided that interest rates didn't adjust that smoothly or completely after all to compensate for the effects of changes in the dollar purchasing power, so stable prices are necessary for a fair and transparent monetary system:

The bimetallists were partially right in their claim that the creditor class were gainers during the period of falling prices in the two decades 1875–1895. The situation has been the exact opposite during the decade 1896–1906. We must not make the mistake, however, of assuming that the enrichment of the debtor-class during the last decade atones for the impoverishment of that class during the previous two decades; for the personnel of social classes changes rapidly. Nor must we make the mistake of assuming that the debtor-class consists of the poor. *The typical debtor of to-day is the stockholder, and the typical creditor, the bondholder.*[84]

Under the prevailing monetary standard, the U.S. dollar was fixed in terms of gold *weight* but not in terms of gold *value* or purchasing power. That guaranteed that the dollar's domestic purchasing power would rise and fall with the supply and demand for money. Most people, even the most sophisticated investors and businessmen, viewed the dollar as a measure of value, and found it difficult or impossible to track or anticipate changes in value. Inflation and deflation were harmful because investors, consumers, and businessmen couldn't predict them perfectly—or even accurately gauge their magnitude in the present and recent past. Decisions made on the basis of faulty expectations necessarily resulted in faulty investment decisions and, from the vantage point of the economy as a whole, too much investment in some areas and too little in others: "reckless wastefulness, for which there must be a day of reckoning in the form of a commercial crisis."[85]

Consider what had happened in a period of sixty years. First, Charles Dickens, Henry Mayhew, and Karl Marx described a world in which the material conditions that had condemned mankind to poverty since time immemorial were becoming less fixed and more malleable. In 1848, Karl Marx showed how competition drove businesses to produce more with the same resources, but argued that no means existed for converting production increases into higher wages and living standards.

Then, in the 1880s, Alfred Marshall discovered that an ingenious mechanism of competition encouraged business owners to make constant, incremental improvements in productivity that accumulated *over time* and, simultaneously, compelled them to spread the gains in the form of higher wages or lower prices, again, over time. As long as productivity determined wages and living standards, people could alter material conditions individually and collectively by becoming more productive.

Beatrice Webb invented the *welfare state* as well as her own vocation as social investigator. Mill argued that a welfare state would eventually absorb the entire tax revenue, and Marx insisted that such a state was a non sequitur. Webb, on the other hand, showed that destitution was preventable and that providing education, sanitation, food, medical, and other forms of in-kind assistance would increase private sector productivity and wages more than taxing would decrease it. In other words, helping the poor become literate, better fed, and less disease-ridden was more likely to raise rather than be a drag on economic growth.

Irving Fisher was the first to realize how powerfully money affected the real economy and to make the case that government could increase economic stability by managing money better. By pinpointing a single common cause for the seemingly opposite ills of inflation and deflation, he identified a potential instrument—control of the money supply—that government could use to moderate or even avoid inflationary booms or deflationary depressions.

Chapter V

Creative Destruction:
Schumpeter and Economic Evolution

Historical development which would normally take centuries
[was] compressed into two or three decades.

— *Rosa Luxemburg,* The Accumulation of Capital, *1913*[1]

On November 4, 1907, news of a run on the Knickerbocker Trust Company in New York set off a stampede on the London Stock Exchange. Frightened investors seeking safety overwhelmed the Bank of England with demands for gold bullion. Threatened with a massive outflow of reserves, the bank responded by raising the interest rate it charged other banks for overnight loans. In the midst of the panic, Joseph Alois Schumpeter, gentleman, and Gladys Ricarde-Seaver, spinster, quietly tied the knot in a registry office near Paddington Station. By the time the discount rate reached 7 percent for the first time in forty years,[2] the newlyweds had left for Cairo, Egypt.

At twenty-four, Schumpeter was already a man of the world. He was born in a small factory town in what is now the Czech Republic, the only son of a third-generation textile manufacturer. After his father's untimely death in a hunting accident at the age of thirty-one, his mother, Johanna, who was and would remain the most important person in Schumpeter's life, resolved to go to any lengths to ensure a brilliant future for her four-year-old son. Largely for his sake, she contrived to move to Graz, a pleasant university town. When her darling turned eleven, she married a retired general thirty years her senior, and convinced her husband to relocate the family to Vienna to a luxurious apartment off the Ringstrasse. Thanks to his stepfather's aristocratic connections, Schumpeter attended an ancient academy for the sons of nobles. At the Theresianum he acquired, in addition to skill in fencing and riding, fluency in no fewer than five classical and modern foreign languages, and invaluable social connections—as well as the flowery manners, promiscuous habits, and extravagant tastes of the titled set. His elite education came at an emotional price. The alter ego of the young social climber was the solitary and driven scholar who read philosophy and sociology texts. In a school where "to be a bit stupid" implied an aristocratic lineage, his brains and obsessive middle-class work habits only underscored his parvenu status.[3] Short, slight, and swarthy, with an unusually high forehead and penetrating, slightly protuberant eyes, his exotic appearance provoked sly taunts about "eastern" (read Jewish) origins. He compensated by excelling in riding, fencing, and verbal wit, and learned to bury his private anxieties under a blasé, ironic, world-weary, manner.

By 1901, the eighteen-year-old Schumpeter had succeeded in graduating from the Theresianum with top honors and gaining admission to the University of Vienna—the first step in what he and his mother hoped would be a rapid climb to the most rarefied reaches of Viennese society. True, Vienna's "first society" was essentially limited to the emperor and his court. But the occupant of a university chair or a cabinet post could breach the "second society," where the clever and capable mingled with aristocrats and plutocrats. By the time Schumpeter was a first-year law student, he already pictured himself as the empire's youngest university professor and the emperor's most trusted economic advisor.

Belle époque Vienna is often depicted by historians as a decadent, complacent, ossified society, and the Austro-Hungarian Empire as hopelessly backward compared with England, France, or Germany. Oszkár Jászi called Austria-Hungary "a defeated empire from the economic point of view."[4] Carl Schorske described the bourgeoisie as politically passive.[5] Erich Streissler bemoaned the dearth of entrepreneurship and the tendency of the sons of businessmen—Ludwig Wittgenstein and Franz Kafka, among others—to choose the arts over industry.[6] In Joseph Roth's 1932 decline-and-fall novel about the Hapsburg monarchy, *The Radetzky March*, a Viennese aristocrat, Count Chojnicki, attributes the empire's seemingly moribund state to the fact that "this is the age of electricity, not of alchemy." Pointing to a brilliantly lit electrical chandelier, he exclaims, "In Franz Joseph's castle they still burn candles!"[7]

In reality, Vienna was infatuated by modernity. As early as 1883, tens of thousands of visitors were being whisked by electric train to the Prater, the vast "people's park" on the Danube, to witness the biggest display of light and power the world had ever seen, the International Electricity Exhibition. Six hundred exhibitors, including America's Westinghouse and GE, Germany's AEG, and Sweden's Ericsson, displayed fifteen accumulators, fifty-two boilers, sixty-five motors, and one hundred and fifty electrical generators. In the "telephonic music room" visitors could "hear the music and singing going on in the opera without moving one step."[8] In another exhibit they could listen to the latest bulletin from a Budapest-based news service for telephone subscribers. The bravest could let themselves be whisked to the top of a 220-foot rotunda, blazing with 250,000 candlepower, in a glass-enclosed hydraulic elevator. At the opening ceremony, Crown Prince Rudolf spoke proudly of "a sea of light" that would "radiate" from Vienna to the rest of the world."[9]

In the race to electrify, Vienna compared favorably to London. Telephone service began in 1881. Trams replaced horse-drawn buses in 1897. By 1906, when the opera *Die Elektriker* opened, the ten inner districts of the city had power. "Elektrokultur" became the slogan of Vienna's entrepreneurs. Every housewife dreamed of an electric hookup that could banish soot and fumes from her kitchen. Factory owners wanted state-of-the-art factories with electric lighting and electric power–driven machinery. Physicians such as Sigmund Freud were eager to try electroshock therapy on their patients. Ludwig Wittgenstein's grandmother took his cousin, the six-year-old Friedrich Hayek, out for a spin in her new electric car.

While it was true that Emperor Franz Joseph spurned elevators and electric lighting, his son, Crown Prince Rudolf, was a staunch supporter of modern industry. Austria had the fourth-largest concentration of commerce and manufactures in Europe, producing steel, textiles, paper, chemicals, and cars. Vienna served as the administrative, trade, and financial center for a vast hinterland that provided the new megalopolises of Europe with food, fuel, and raw materials. The economic upswing of the late 1870s through the mid-1880s had created a boom in exports of sugar and textiles as well as in railway construction. By the late 1880s, electrification had displaced the railroads as the principal magnet for new investment.

The city's architecture reflected not only imperial but also bourgeois aspirations. The Ringstrasse, the wide boulevard that encircled the inner city with its neoclassical parliament building and baroque opera house, interspersed with the mansions of the "Boulevard Barons," reflected the stupefying progress of the times. The luxury rental apartment, or *Mietpalais,* rather than the villa, attracted the nouveau riche, the parvenu, the social climber. Middle-class, multiethnic, and determinedly monoculturally German, Vienna was the destination of choice for refugees from the rest of the empire—especially after 1867, when liberals in the cabinet promoted Jewish emancipation along with economic modernization. Many of the recent immigrants became peddlers or small shopkeepers. Their sons mostly went into professions such as law or medicine that didn't require attendance at an elite prep school, or banking, journalism, or the arts where a university degree was not needed. The preponderance of Jews in law, medicine, banking, journalism, and the arts stoked resentment, especially in bad times. As one historian put it, "Anti-Semitism rose as the stock market fell." [10]

Economic data contradict the stereotype of economic decadence. Not only did the economy grow three times as fast between 1870 and 1913 compared with the previous forty years, but per capita income doubled in real terms in that period, even as the population surged. True, Vienna suffered the same chronic shortage of housing, sewers, clean drinking water, and paved streets as Victorian London. But the economic historian David